Genomes and What to Make of Them

Genomes and What to Make of Them

Barry Barnes & John Dupré

The University of Chicago Press :: Chicago and London

Barry Barnes is a codirector of the ESRC Centre for Genomics in Society at the University of Exeter, at which he was formerly professor of sociology. He is the author of several books on the sociology of the sciences and was awarded the J. D. Bernal Prize for his career contribution to the field. **John Dupré** is the director of the ESRC Centre for Genomics in Society, professor of philosophy of science at the University of Exeter, and the author of several books, including *Darwin's Legacy: What Evolution Means Today.*

The University of Chicago Press, Chicago 60637
The University of Chicago Press, Ltd., London
© 2008 by The University of Chicago
All rights reserved. Published 2008
Printed in the United States of America

17 16 15 14 13 12 11 10 09 08   1 2 3 4 5

ISBN-13: 978-0-226-17295-8 (cloth)
ISBN-10: 0-226-17295-3 (cloth)

Library of Congress Cataloging-in-Publication Data

Barnes, Barry.
    Genomes and what to make of them / Barry Barnes & John Dupré.
        p. ; cm.
    Includes bibliographical references and index.
    ISBN-13: 978-0-226-17295-8 (cloth : alk. paper)
    ISBN-10: 0-226-17295-3 (cloth : alk. paper) 1. Genomics. 2. Genomes.
I. Dupré, John. II. Title.
    [DNLM: 1. Genomics. 2. Genome. QU 58.5 B261g 2008]
    QH447.B38   2008
    572.8'6—dc22

                                        2008024432

Contents

Acknowledgments

The first debt we are happy to acknowledge is to the Economic and Social Research Council (ESRC) UK, without whose generous funding of the ESRC Centre for Genomics in Society (Egenis), the research center of which we are directors, this book would not have been written. The intensive study of genomics and its social consequences that has taken place there over the last five years is reflected throughout the book. We are also indebted to the Arts and Humanities Council (AHRC) UK and to the Wellcome Trust for grants in support of research at Egenis that we have drawn upon; and to the University of Exeter for its uncompromising support of our establishment and running of the center. We want to express our gratitude for this both in general and more specifically by thanking those, including Ajit Narayanan and John Bryant, who helped to get our project under way, and most of all our Head of School, Jonathan Barry, who has been helping us in all kinds of ways ever since.

Our greatest personal debt is to the third director of Egenis, Steve Hughes, who has brought to the center a lifetime of experience as a genomic scientist and an inexhaustible willingness and talent for explaining biological detail. We are just two of the many to have benefited from Steve's "technical briefings" at Egenis, and the limited grasp of the field we have managed to acquire as outsiders owes a great deal to him. We need also to thank the large number of colleagues in Egenis who have made the center an extraordinarily stimulating place to work and whose inputs, in many instances unacknowledged, are in evidence throughout the book. We especially need to mention the work of Maureen O'Malley on systems biology and microbiology that proved indispensable in the writing of chapters 3 and 4, and to thank Adam Bostanci, Jane Calvert, Jonathan Davies, Christine Hauskeller, Ingrid Holme, Mario Moroso, Staffan Mueller-Wille, and Alex Powell for helping our writing along in even more ways than they may be aware of. Another person without whom the book could not have been written is Cheryl Sutton, the administrator of Egenis, without whose remarkable organizational skills we should have lacked the time to write. We would also like to thank the various people who have over the years made up the rest of the Egenis administrative team, all of whom have contributed to making this a pleasant and productive place in which to work.

We need also to thank many other academic colleagues both at Exeter and elsewhere for their support and inspiration. Particular thanks are due to David Bloor, Philip Kitcher, and Hans Jörg Rheinberger. The thoughtful comments of anonymous reviewers for the University of Chicago Press were also of great help to us and are gratefully acknowledged. Finally we, and JD in particular, want to thank Regenia Gagnier. Not only have John's ideas, as always, benefited throughout from her penetrating criticism and exceptional synthetic insight, but she has constantly if unintentionally reminded him that there is more to life than ideas.

Introduction

A few miles outside Cambridge, England, in the small village of Hinxton, are the sprawling grounds and buildings of the Wellcome Trust Genome Campus, home to the Sanger Institute. The first thing the visitor to this iconic site of modern biology will see, in a bold electronic display above the reception desk, is a rapidly passing sequence of C's, G's, A's, and T's, which, she will be informed, constitute a real-time readout of DNA that is being sequenced somewhere on the premises. The first stop on a tour of the building is a room in which large robots stick tiny probes into Petri dishes and then into rectangular arrays of test tubes. Spots on the nutrient gel in the Petri dishes, she will be told, contain bacteria infected by viruses with fragments of human DNA. The replicating bacteria are the means of replicating the human DNA within to produce whatever quantities are required. Next, the visitor is invited to peer through a window in the closed door of a room containing the small but expensive machines that perform the polymerase chain reaction, a process that can selectively amplify minute amounts of specifically targeted lengths of DNA. Finally, there are the sequencing machines themselves, occupying a warehouse-sized space in which conversation is made difficult by the hum of a powerful cooling system. There are perhaps a hundred sequencers, looking somewhat like large white microwave ovens, and each is connected to a familiar-looking desktop computer. The room is almost entirely devoid of humans, and the process is almost wholly automated; a solitary technician delivers a tray of material to one of the machines.

It is difficult to imagine behind this routine activity the many years of creative work that made this room possible, or to appreciate the importance of what is going on in it and in the instruments it contains. The visitor is in the presence of the powers of a new technology, but its proximity and visibility are of little help in grasping what those powers amount to. And not even its creators and controllers know how the technology will be employed in the future or how it will be developed still further to provide still greater powers. All they can be sure of is the past, and the great purpose that led to the assembly of this extraordinary array of instruments and people. It was at Hinxton and a small number of similar locations around the world that the sequencing of the human genome was carried out.

The publication of the first draft sequence of the human genome at the beginning of this millennium was a grand ceremonial occasion. Leaders of the project joined the president of the USA and the prime minister of the UK in marking the event; the appropriate platitudes were uttered; tributes were paid and thanks were given. Press and media competed to convey something of the awesome significance of what had occurred, unfazed, as ever, by any uncertainty about what that significance was. Yet the genome project and all that it symbolized had not displayed its significance in quite the way that the only comparable project in our entire history had done. The Manhattan Project had announced its completion with explosions at Hiroshima and Nagasaki, leaving no doubt that the world was changed thereby. Many struggled in the succeeding decades to come to terms with what had happened, not least some of the participants in the project itself. Things were rather different sixty years later, when publics still less than fully clear what the rejoicing was about were soon overwhelmed with different preoccupations by explosions of another sort. Even so, those who made so much of the project were not wrong: it, or rather that of which it is a symbol, may transform the world as profoundly as the Manhattan Project and what it symbolized. We are now living with a new knowledge and a new technology just as potent as that of nuclear physics and engineering, albeit one whose powers are distributed in very different ways. And many people are struggling to come to terms with this, more even than turned to the problems and possibilities represented by nuclear fission in 1945.

Developments of this sort never arrive out of the blue, of course. New knowledge and technology, like new organisms, arise from the modification of older forms, and must be understood in part in terms of their ancestry. The emergence of genomics has diverse and complex interrelations with the genetics that preceded it and is a central part of a larger transformation in which not only genetics but large parts of the life sciences generally are being "molecularized." This process continues apace and is now bringing us into an era in which proteins and other biological macromolecules are being investigated at the same level of detail as the DNA molecules that constitute genomes. And indeed the transition to genomics described in what follows is best addressed as a microcosm or synecdoche of this larger transformation, and the sequencing of the human genome as a symbol of its progress. Even so, our account speaks almost entirely of a transformation in genetics and of the emergence in the second half of the twentieth century of a new molecular genetics and eventually of genomics. Although the extent to which traditional genetics has been replaced, transformed, or superseded by genomics and other branches of contemporary molecular biology is a question that admits of no simple answer, we shall emphasize the contrast between genomics and traditional genetics, partly for brevity, but more importantly because we seek to highlight difference and its consequences. Crucially, we want our story to emphasize the profound change that the rise of genomics is making to our powers and capacities.

Classical genetics as it unfolded in the first half of the last century was one of the most successful of the biological sciences. It transformed our understanding of inheritance and, eventually, in conjunction with the theory of natural selection, our understanding of the evolution of species as well. And it did this using a remarkably modest range of methods, procedures, and technical tools, through one of the most impressive programs of careful, laborious, patient, and well-organized research in the history of science. It was a program involving the repeated sorting and selective breeding of individuals in large populations of organisms. By systematically recording observable differences in the characteristics of these organisms it was possible to attribute invisible genetic differences to them, and these in turn could be used to account for, and often to predict the distribution of, the same differences in subsequent generations. It cannot be emphasized too strongly that classical genetics was concerned with the investigation of difference; its considerable success lay in predicting and accounting for difference. It did not aim or claim to explain how organisms came to be the way they were, only how they came to differ one from another and, more specifically, how the proportions of organisms with different characteristics changed from one generation to the next. For all its success, however, its methods were mainly observational and its powers to intervene in biological systems and to modify them from the inside remained very limited. It was only in the second half of the century that this crucial limitation was overcome, and it is precisely its overcoming with the development of molecular genetics and genomics that makes the emergence of these fields of such extraordinary interest and the sequencing of the human genome a symbolic event of world-historical importance.

In the period from the elucidation of the structure of DNA in 1953 through the development of recombinant DNA techniques in the 1970s, to the genome sequencing projects of the 1990s, new methods, techniques, and research instruments proliferated, and powers of quite extraordinary scope and sophistication were acquired for changing the structure and functioning of organisms. What had previously been regarded, and not infrequently attacked, as a science offering at most an understanding of the world was being transformed into one with an awesome capacity to change it. It is perhaps as well that this change did not occur earlier, before the war, when many societies were permeated with eugenic enthusiasms and lacked the memories that we now have of where these can lead. Even now, however, the new powers of genomics elicit anxiety as much as hope, and efforts to come to terms with them are permeated at all levels by awareness of how double-edged they may prove to be. Whereas critics of the earlier genetics tended to disparage its ability to explain and predict, let alone to control, both advocates and enthusiasts of the new genomics now agree that this science could come to have truly profound consequences throughout the wider society. They tend to agree as well that among these consequences are some that are completely appalling, and to differ only on how far human beings are capable of avoiding such consequences once technical constraint is overcome and it becomes for the first time a matter of choice whether to do so.

As will now be clear, one of the main objectives of this book is to convey some sense of the new powers with which we are being endowed as genomics and a new molecular genetics grow out of traditional genetics, and of the likely impact and significance of those powers and their use. But alongside that objective lies a second, equally important one. Knowledge, in the sense of theories, schemas, and ideas, has also changed in this context, along with powers and techniques, and we want to convey something of the change involved. We shall be concerned not only with knowing *how*, in this book, but with knowing *that*. There are, however, some difficulties that must be faced here. We are talking of fields in which research is proceeding with unprecedented rapidity, so that whatever we say of the current state of knowledge within them will have to be read as ancient history, setting out findings and theories that could well have been discounted by more recent work. And in attempting to convey the knowledge of a specialized technical field to nonspecialist audiences we will be expected to simplify the knowledge and yet not to distort it or to gloss over important subtleties and complexities, which strictly speaking is impossible. Of course, this last difficulty faces anyone who would write a book of this sort, and a compromise is always necessary between the need to communicate at all and the need to communicate accurately, but a very great deal can nonetheless hang on getting the balance here right.

The briefest of backward glances suffices to establish the importance of this last problem in the context of genetics. Throughout its history genetics has been addressed in terms of simplistic stereotypes, which have encouraged misunderstandings not only of the field itself but of its moral and political significance. We can see this particularly clearly in the period of the early development of the field, when questionable versions of human genetics were appealed to as rationalizations of the noxious eugenic policies then favored by governments in Europe and the USA. But in truth the dangers of simplistic misrepresentations of genetics are apparent everywhere in its history and remain in evidence today. Today, of course, in a post-Holocaust world, the dangers are different, and the need for adequate understanding is engendered by new problems and anxieties. States are now much less inclined to stigmatize and degrade whole groups of their citizens. Memories of the old eugenic programs elicit a sense of shame, and genetic arguments that might conceivably be used to support such policies immediately attract intense suspicion. The need for a more widespread general understanding of genetics/genomics now arises from a technology that seems to offer new possibilities for individual choice rather than resources for the policies of a coercive state. Even so, this constitutes a change of emphasis within, rather than a complete rejection of, the eugenic ideology. The choices becoming available to individuals involve both a positive eugenics of manipulation and "improvement" of the genome, and negative eugenics involving, for example, the genomic examination of embryos and the discarding of "unfit" specimens. Both these possibilities are attracting increasingly intense debate.

Thus, there is yet another strand that needs to be woven into the texture of this book. Genomics/genetics confront us not just as practice and technology, and as knowledge and understanding, but as myth and ideology as well. And just as technology and knowledge systematically change over time, so too do the associated myths and ideologies. It is an attractive vision that the aim of a book such as this should be to separate the myth and ideology from the authentic knowledge of genetics and genomics. But as the intimate entanglement of genetics and eugenics in the early years of this century suggests, this is very probably an impossible task. Myth and ideology can be found not only among practitioners of classical genetics, but also in the historical frames through which it has been represented. Historical accounts have often concealed the diversity of perspectives from which classical geneticists approached their subject matter, and how aware they were of subtleties and complexities that are now sometimes seen as the revelations of more recent and sophisticated science. Indeed, the "heroic history" offered as useful background in our first chapter follows an established pattern that expediently exaggerates the "Mendelian" perspective in genetics and glosses over a good deal else. And more generally, the problem of distinguishing the old genetics from our current myth of it may well now be more difficult to solve than that of distinguishing it from myths contemporary with it. At any rate, there is a need for a major shift in everyday understandings of both classical genetics and contemporary genomics and in the stereotypes routinely deployed to characterize the objects and processes they study, and a major aim of this book is to help that shift along. Even if this were merely a matter of reorienting everyday understandings around a new mythology, it would stand as a worthwhile project.

: : :

We are now in a position to review the contents and organization of the book. As the table of contents makes clear, the chapters cover diverse topics, and it is possible to read some of them as independent units or to focus selectively, whether on the early chapters about the science or the later ones about its impact. Certainly, not everyone will want to devote the same amount of time to the background material provided in chapter 1. The chapter provides a broad-brush picture of genetics and the beginnings of genomics both as knowledge and as activities and an initial description of some of the ideas, theories, methods and procedures, instrumental resources, and experimental systems that will be referred to in subsequent chapters. For some it may serve as a reminder of what they know already, for others as a sketch that can be filled out by the more detailed accounts now widely available both in the bookstore and on the Internet. This is as much as it seeks to do.

Among its limitations, however, one in particular does merit special mention, since it might be seen as a flaw that runs through the entire book. In neither genetics nor, despite appearances, in genomics, has work on human beings figured large at the cutting edge of research. Phages, bacteria, plants, and insects have figured prominently at this level for good technical reasons, to the extent that we have learned as much or more about our own genes and genomes from the study of these kinds of organism as from the study of ourselves. Similarly, the practical use of the knowledge and competence these fields have gained is furthest advanced in the realms of microorganisms and plants, and direct application to human beings is very much a sideshow at the present time. But outside the technical work in these scientific fields themselves, much of the interest in genetics and genomics is focused on their potential applicability to humans, and we have accordingly allowed our simplified story to give a not entirely justified prominence to human beings and human genomes.

Having set the scene, we turn to the first major task of the book in chapter 2, which presents the transition from genetics to genomics as a shift in our knowledge; that is, a shift in concepts, theories, and representations. Commentators have rightly pointed to the growing importance of the physical sciences in genetics, especially after the elucidation of the structure of DNA and the rise of molecular biology, and have seen this as a continuing incursion leading to a new molecular genetics and to genomics. The shift here has often been characterized as a move to reductionism and mechanism that clashes with the distinctively holistic and functional approach of biology, but it is actually better described as a move towards materialism and the molecular/structural modes of representation typical of chemistry. This is a move that threatens neither holistic nor functional forms of scientific theorizing, although it remains of profound importance in other ways. The identification of DNA as the "genetic material" and the subsequent disclosure of its stereochemical structure was initially taken as a vindication of classical genetics and particularly of the materialists among its practitioners. Genes really did exist, it was thought to show: they were segments of DNA. Subsequent work in molecular genetics saw a proliferation of models and theories, among which the most valuable were indisputably the informatic models that famously treated DNA as a four-letter text and inspired the efforts to "read" or "sequence" it in its entirety in different organisms that subsequently grew into genomics. But the different kinds of models now employed in research—classical, informatic, and stereochemical/molecular—came increasingly to diverge, and in particular it became more and more awkward to maintain compatibility between representations that referred to genes and representations using models of DNA. And when the question was raised as to what was really going on or what was really being studied, few doubted that it was molecules doing their thing rather than genes doing theirs, given that it had to be one or the other. Ontological authority shifted to molecular/stereochemical models, and it began to make sense to say that there really are genomes, but not genes.

The simplest and most dramatic way of characterizing the transition to genomics is to point to this shift, but of course other changes were also involved. Thus, classical genetics had framed its problems in terms of inheritance, and sought to account for manifest differences between organisms in terms of invisible genetic differences passed on from ancestors to descendants. But its successor sciences have become more interested in the functioning of genomic DNA throughout the entire life cycle, as a part of a larger chemical-molecular system located within the cell. Indeed, in textbooks on the new sciences the older genetics is now often referred to as "transmission genetics" and treated as just one of the many specialized subfields of the discipline as a whole: sometimes hundreds of pages pass before it is even mentioned. Nor are these sciences concerned with difference to the extent that the old genetics was, and had to be. The methods and techniques now available to molecular geneticists allow them to study DNA in the laboratory as one ingredient in the functioning and regulation of cellular chemical processes, with the aim not primarily of explaining differences but of throwing light on the "normal" functioning of cells and organisms. And in genomics a whole range of mathematical methods and techniques are now used to analyze DNA sequences stored in computers, sometimes for the purpose of making comparisons between organisms, but often in the hope of inferring the normal function of the DNA from its sequence. All this is a far cry from traditional genetics.

To summarize, the transition that concerns us has involved genomes rather than genes being treated as real, and systems of interacting macromolecules rather than sets of discrete particles becoming the assumed underlying objects of research. Inheritance and inherited differences have ceased to be central, whether as subjects of investigation or as schemas that frame perception, and the life cycle has replaced descent as the dominant frame within which to order experience and turn it into knowledge of living things. Finally, although this important change is mentioned rather than discussed in chapter 2, research itself has increasingly involved intervention and control as well as observation and selection and both the in vitro laboratory methods of molecular genetics and the in silico mathematical methods increasingly employed in genomics stand as new and significant departures from what had gone before.

We now have in place the necessary background against which the more general and wide-ranging overview of genomics in chapter 3 can unfold. We begin with a discussion of what genomes are, or more precisely of the various ways in which scientists refer to them and seek to define them. So far we have spoken of genomes as the DNA molecules sequenced by researchers in enterprises like the Human Genome Project (HGP), but they are actually understood and defined in a number of different ways. Indeed, in a different kind of book the different ways of defining genomes could be used to illustrate what profound effects researchers' methods have on their understandings of the objects they investigate. Addressed via the methods of molecular genetics, genomes are made visible as DNA molecules. For those who deploy mathematical methods, genomes are liable to be perceived as long four-letter sequences virtually constituted in computers. Maximum continuity with the methods of classical genetics is sustained if genomes are defined as inventories of the entire genetic material of an organism, containing all of its genes. And another valuable link with older methods and practices can be made by defining a genome as all of the DNA in a single set of the chromosomes of an organism. But while specialist researchers have to conceptualize the objects they study in ways related to the methods they use to study them, those looking in from the outside have to do the opposite and seek a method-independent understanding of scientific objects. Any such understanding is bound to have its limitations, but our argument in chapter 3 concludes that as a basis for everyday understanding, genomes are best thought of as material objects made of DNA, and that scientists should be understood as referring to such objects when they speak of genomes.

We are then able to go on to discuss what genomes are like as objects, touching upon their shape and internal structure and how they vary, and stressing especially how very much more is found in (most of) them than "coding" DNA. We also include a discussion of epigenomes, which create problems for the view of genomes as molecules in organisms, since epigenomes are in truth those molecules, and of epigenetics, which constitutes a fascinating enrichment of our understanding of inheritance, of ontogeny, and of the chemical functioning of DNA in the cell. In fact, as we explain, the four-letter sequences described in chapter 2, despite their immense importance, turn out to embody a considerable abstraction from what is now seen as the material reality of genomes, and the finer level of detail described by epigenetics is essential if we are to know what particular genomes in particular cells are disposed to do.

Any new science identifies new similarities and differences between things that can be used to classify them and order them into taxonomies. That this is indeed the case with genomics will become clear as the book proceeds, and specific claims about genomic similarities and differences are encountered, but in anticipation of these chapter 3 discusses some important general problems associated with similarity and difference and gives warning of a need for caution in the face of claims of this sort. It is never wise to accept without further examination unqualified claims that one thing is similar to or different from another, or even that they are 97% similar, or 99% similar—which is the kind of claim that is increasingly being made about the genetic/genomic similarities that exist between different living things. Whether it is assertions of the genomic similarities between different humans or groups of humans that are concerned, or those between humans and other animals, or even those wholly between different animals or animal kinds, it is important to remember both that they are often prompted by expediency and illicit objectives and that they are always open to question and contestation. Between genomes, as between the organisms they inhabit, and indeed as between objects of any sort or kind, there are at once no "natural" measures of degree of similarity and far more empirically recognizable similarities and/or differences than can ever be taken into account. As a consequence, difficult issues surround questions of genomic similarity and genome-based classifications, and some appreciation of these difficulties is essential to a properly critical assessment of what can be inferred from the findings of genomics, just as it was in the past where the findings of the old genetics were concerned.

Chapter 3 concludes with a brief description of two of the main directions in which research is currently moving. Genomics has produced almost unimaginable quantities of biological data, but much less certainty as to what can be done with it all. Synthetic biology and systems biology can be seen as attempts to reduce the complexity represented by this vast amount of data, with the former mainly involving an extension of existing laboratory methods in an attempt to produce simple forms of life or components of living things in vitro, and the latter using mathematical methods to construct models of the main features of complex molecular systems in silico. We are in no position to say how successful either of these two lines of research will prove to be, but as we write, major technical achievements are being claimed in both of them that could, if substantiated, offer striking confirmation of the utility long expected to ensue from genomics research.

With the discussion in chapter 3 as background, and genomes having now been discussed at some length as molecules, chapter 4 is able to turn to the importance of genomics in the broader context of biology and in particular in relation to the theory of evolution. We approach evolution, however, from a direction foreshadowed in the previous chapter, by first considering how the classification of organisms has been transformed by the rise of genomics and the systematic comparison of DNA sequences in the genomes of different organisms that it has made possible. As elsewhere in biology, the essential framework in which such comparison occurs is evolution, and evolutionary genomics is now an important field in which the phylogenetic relations between genomes, including our own, are systematically explored. Interest in the number of genes in the human genome, in the complexity of its organization, and in its similarities and differences with other genomes, whether those of our fellow primates or those of far simpler organisms, have figured large among the products of this interest, inspired in some cases by a desire to be reassured of our special position as a species within a larger natural order of species.

Although questions of this last sort are often framed in ways that place them outside the proper realm of biological (or any other) science, there are ways in which genomic research can still be deployed in opposition to them, as contributions to naturalistic accounts of evolution that radically decenter not only the human species but even the very idea of a division of organisms into species. We emphasize the virtues of accounts of this sort in our discussion, and in particular we promote accounts that focus attention on the evolution of molecules and of microbes. Most standard accounts of evolution, and certainly the great majority of those written for nonspecialist audiences, focus very strongly, sometimes entirely, on complex multicellular organisms and particularly on the place of humans among them, whereas a properly comprehensive account ought to begin at the beginning with genomes and the associated molecular systems and move on to give heavy emphasis to the microbes that were the sole living things for the greater proportion of evolutionary time and still constitute the greater part of the biomass of the planet today and carry most of its evolving genetic material.

One exciting consequence of the development of the contemporary tools of molecular biology is that they have enabled us to investigate in unprecedented detail the microscopic organisms that, as we have just noted, are by far the commonest living forms. We conclude chapter 4 by pointing to some remarkable ways in which genomic research on microbes is transforming biological ideas. In discussing classification and species we note how, at least among microbes, genetic material passes with unexpected ease between organisms of different kinds, problematizing assumptions about the distinctiveness of kinds and the processes of evolution. The extent of this circulation of genetic material, combined with the realization that microbes generally exist in complex interconnected communities in which DNA circulates quite freely, has led to speculations about a genetic commons, and a developing field of "metagenomics." The implications of this are not limited to the understanding of microbes, however. Macroscopic organisms such as ourselves live in symbiotic relationship with vast colonies of microbes, and the developments we are describing here threaten not only the central place of humans in the biological world, but even our intuitive understanding of what constitutes a distinct biological individual.

Chapter 5, the last chapter in which we are content to discuss genetics and genomics largely as forms of knowledge, is very different in its emphasis from those preceding it. Its primary focus is on the ways in which genetics and genomics have been deployed as resources for explanation. It looks backward to classical genetics, and outward to the ideas of genetics and genetic explanation that exist beyond the field, among nonspecialists. And it concentrates almost entirely upon human traits, especially human behavioral and psychological traits, not because these are the traits that geneticists have been most intent upon explaining, but because it is those traits, and the possibility of explaining them in "hereditarian" terms, that have most interested outsiders. To understand this emphasis we need to return to the gene concept and its recent career. When in chapter 2 we identified a shift in ontological authority from gene to genome, we did not wish to imply that the former term was headed for the waste bin. Indeed, it is likely to continue to do valuable technical work, both in exchanges between specialists and in those involving the communication of genetic information to outsiders who need only the gist of it, carriers of disease susceptibility genes, for example, who may be more interested in identifying a compatible partner for sexual reproduction than in the finer details of molecular structure. Classical genetics, built around the gene concept, can offer some simple and memorable explanatory patterns of great utility. But unfortunately these patterns have been further simplified and reified, and the crucial fact about them—that they are explanations of differences by differences, that they offer understanding not of traits or characteristics but of variations in traits or characteristics—has been ignored. The result is what we refer to in the chapter as "astrological genetics," a crude form of genetic determinism currently thriving in the media and the wider culture, which has proclaimed the discovery of "genes for" intelligence, homosexuality, aggression, and a succession of other human behavioral traits selected with a mind to the interests of the general reader. This form of discourse, which sports a pedigree as long as or longer than that of genetics itself, is of course commonly encountered in moral and political debates in which the existence of "genes for" this or that, and their distribution in society, are alleged to justify policies and attitudes of various sorts, not infrequently noxious and repugnant.

The history of the political and moral use of genetic and pseudogenetic hereditarian arguments is so well known as to need no more than the briefest mention here. It needs to be alluded to only as a reminder of how undesirable is the stubborn persistence of misconceived forms of genetic determinism, and a thriving astrological genetics that happily speaks of genes for practically every human behavioral trait you care to name, without any regard for the fact that genetic causation, if it exists at all, can exist only in a specific range of environments, operating in conjunction with further causes located in those environments. In fact, accounts of the flaws of astrological genetics and genetic determinism, and analyses of what can and cannot be asserted by way of hereditarian explanations of traits and their distribution, have been offered again and again by classical and population geneticists themselves over the years, but no amount of careful and authoritative commentary has succeeded in eliminating the misconceptions constantly recycled in these contexts. Controversies between hereditarian and environmental explanations of variations in behavioral traits in human populations continue even now to add to what is an appalling tangle of reputable argument and astrology.

The final section of chapter 5 asks whether a genetics/genomics is possible that could offer plausible explanatory accounts of human behavior without taking the road into this morass. One possible alternative here is to construct explanations that follow the long route from the human genome and its setting, via proteins, cell chemistry, cell differentiation, the drawn-out growth of the brain and nervous system, and so on to the behavior that is its intended destination. The daunting problems that must be faced in following such a long and complex path through a veritable maze of causal connections are, however, obvious, and those who have chosen to set out upon it, it has to be said, have so far traveled hopefully but have yet to arrive. Even so, there can be little doubt that causation does move along this route, although not necessarily as one-way traffic, and that explanatory insights of some sort will eventually emerge from their research.

Chapter 5 ends by suggesting that widespread interest in genuine genetic/genomic explanations is likely to emerge only as and when the powers of the new sciences are widely deployed, and in chapter 6 we specifically discuss genomics and genetics as power. With this chapter the primary focus of the discussion finally shifts from genomics as knowledge to genomics as people doing things, and offers a picture of a field in which powers are accumulating at an unprecedented rate as the skills and techniques of the physical sciences flood into it and fruitfully combine with its existing practices. Even so, not everyone acquiesces in this widespread vision of a burgeoning technology that is going to enable us to do more or less what we wish with genomes, and indeed there are good reasons for doubting it. The hype surrounding the Human Genome Project, which emphasized the improvements that would follow for our health and well-being, has engendered a sense of anticlimax. "Where are the promised benefits?," it is asked, and even "What has it contributed to science?" For indeed, relative to the vast outlay there is little yet to show by way of benefit, whether economic or intellectual, from a project perhaps better understood as an attempt to boost the infrastructure of a new technoscience than as a response to the need to read every last letter in the book of life. Nor do we doubt that in a larger sense there is currently little to show for all the work done, at least in terms of controlled interventions into our genomes and the systems in which they function. But in our view this is but the calm before the storm, for which regulators and political authorities ought perhaps to be grateful. And if we look beyond the narrow confines of the HGP and the professed aspiration to manipulate the human genome for medical reasons, things are already very different.

Intervention and control involving nonhuman genomes, especially plant genomes, are still haphazard, but they are of great economic importance nonetheless and developing rapidly. And in relation to human genomes, techniques of scrutiny and diagnosis are evolving and rapidly finding applications even if techniques of intervention are not. Indeed, the feature of genomic powers that currently needs the most urgent consideration is arguably the way that they facilitate the application of *other* powers and capacities. The new genomic technologies are currently amplifying the existing powers of bureaucrats and administrators to a worrying degree, the more so as the IT increasingly central to both genomics and bureaucracy proliferates and becomes ever more powerful, and data on the genomes of ever greater numbers of people are committed to disc for one reason or another. The prospect of death by administration as a side effect of all this should not be lightly dismissed, although of course many other potent forces are simultaneously pushing us in this direction.

Having paused to consider this prospect, the discussion moves back to genomic powers themselves and the debate and controversies that increasingly surround the prospect of their use. The commercialization of genetically modified organisms, especially as food crops, continues to arouse passionate controversy in Europe and is currently being strongly resisted by a number of activist groups. And insofar as the human genome is concerned, the possibilities of cloning, of biological parentage by both members of a homosexual couple, or even by multiple "parents," of gene therapy, of replacement tissue and organ manufacture, of DNA-specific biological weapons, of individual humans or even entire human populations modified to produce a eugenic Utopia, are just a few of the scenarios debated in anticipation of what might lie just around the corner genomically. One thing that these debates reveal is that the ever-increasing powers and capacities that genomics and related sciences are offering to us, far from being everywhere regarded as unmitigated goods, are uncompromisingly opposed by substantial social constituencies as intrinsic evils, and that their use is rejected still more widely as involving risks and dangers not worth tolerating for the sake of what appear to many to be no more than marginal benefits. A crucial problem here, which we discuss at some length, is that of accounting for the very different perceptions of the uncertainties, and hence the dangers, associated with the use of genomic technologies. These differences in perception seem to be best understood by reference to social and cultural differences between the groups of people involved rather than by their differential access to information or differences in their capacity to calculate risk. Variations in perceptions are evidently socially structured, although how to develop this into a satisfactory explanation of them remains moot. One conjecture frequently put forward here is that technologies are trusted or distrusted to the extent that the institutions that carry and deploy them are, and that the preexisting orientations of people to institutions are the key to an understanding of what is going on here. We support this conjecture by noting how many features of the recent controversy in the UK over the safety of genetically modified (GM) foods are consistent with it, and make brief mention of the quite different reception of these products in the US to further increase its plausibility.

Chapter 6 discusses genomic powers almost entirely in a utilitarian frame, but much of the resistance to their use is actually rationalized in nonutilitarian terms. Chapter 7 looks at some of the major forms of non-utilitarian argument that have been deployed here. It gives particular attention to claims that genomic interventions undermine natural order and harmony, and even more to arguments that indict them as threats to human dignity. And indeed these last deserve to be dwelled upon, since they have been particularly prominent in recent debates, and philosophi-cal formulations of them, far from being mere academic abstractions, have significantly informed and influenced legal and regulatory inno-vations in this context. The work of the philosopher Jürgen Habermas and his arguments against the deliberate "reprogramming" of the hu-man genome provide good illustrative material here, although we don't claim to resolve the controversial issues with which his arguments are concerned. Indeed, our own concern is to stress that reasonable human beings with different backgrounds or in different situations may find arguments of this sort more or less persuasive, or even more or less in-telligible, and that complete agreement on these kinds of issue is not to be expected. There will be endless argument about how far genomic in-terventions threaten dignity and good order just as there will be about how far they threaten our safety. These issues will not be settled by non-utilitarian argument in the former case any more than by utilitarian ar-gument in the latter. In chapter 6 it was claimed that people in different contexts assess the threat to our safety posed by genomic manipulation very differently, because of different prior conceptions of the uncertain-ties associated with such manipulation. The assessments of threats to our dignity discussed in chapter 7 must be expected to vary similarly, because of different prior conceptions of human dignity, of how far it extends to genomes, and of what counts as an insult to it. And in both cases, explanations of these differences in prior conceptions are likely to be found in the different social locations and institutional orientations of those who hold them.

Chapter 8, as well as reviewing the overall argument in order to make an ending, also seeks to evaluate some of the ways in which we might attempt to solve the problems created by the extraordinary scien-tific and technological developments with which the book has been con-cerned. Time and again as the argument has proceeded, there has been cause to note the exceptional status attributed to genetics and genomics. Whether because of the exceptional objects they study, or the exceptional powers they confer or the exceptional risks associated with them, they are given a different status from other bodies of natural knowledge, and presented, justified and, above all perhaps, criticized in specially constructed frames. Only nuclear physics and nuclear weapons technology have been accorded similar treatment in recent times, and the reason in both cases is surely the same. Both are sources of truly exceptional powers and hence of exceptional opportunities, and by virtue of this they are also foci of exceptional fears, fears of the capacities they represent in themselves, and fears of each other as agents able to wield those capacities. And of course the fears are justified in both cases, and so to some extent are the exceptionalist orientations they inspire. Here indeed, in the unbridled powers proliferating within it, lie the true problems with which genomics confronts us, problems that demand of us the courage either to ride the tiger we have created or to admit that this is something beyond us and to recognize the need for renunciation.

Unfortunately, however, naturalistic orientations of this sort easily metamorphose so that nonnaturalistic forms of exceptionalism come into existence that are far more seriously flawed and need to be called into question. Some of these derive from genetic, and now genomic, essentialism, wherein genomes, and above all human genomes, are assumed to embody the natural species essences of the corresponding organisms, and in the case of human genomes the specific individual essences of particular human beings as well. This genomic essentialism encourages critics and commentators to develop nonnaturalistic interpretations of genomics and its objects of study, focused in practice almost exclusively on human genomes, and from this many different forms of exceptionalist thinking can take off. The astrological genetics/genomics discussed in chapter 5 is one example of this. Habermas's account, discussed in chapter 7, in which our genome is spoken of as our "natural essence," is another, and there are further variations on this essentialist view that would extend attributions of human dignity back through the fetus and the embryo to the zygote and its genome(s), to apply throughout the entire human life cycle, on the basis of contested natural analogies and even more contested claims to religious authority.

Genetic/genomic exceptionalism of this kind is often encountered as a part of a more general tendency to reify human nature and to misrepresent it in the form of a material object. Reification of this particular sort is unnecessary in theory and tends to be harmful in practice. And the historical precedents strongly suggest that this will continue to be the case if we assume that our essential nature is constituted by, or indistinguishable from, or embodied in the material objects we refer to as genomes. Whether it is ever wise to speak of "human nature" and thereby to risk hypostatizing it is moot, but in any case our discussion points to more promising ways of conceptualizing this problematic entity than those that treat it as incarnate in genomes or indeed in material objects of any sort.

It remains arguable, of course, that genomes should not be regarded as material objects at all and that a better way of undermining genetic/genomic essentialism is to regard genomes (like genes) as theoretical constructs and not as things out there in the world. But if only because human beings understand each other most easily, clearly, and effectively in a realist frame, our own preference has been to speak of genomes as real, material objects while taking care to avoid the problems associated with most existing forms of genetic/genomic realism and essentialism by use of a different and less compromised ontology. The decline of the ontological standing of the gene concept among specialists, along with the shift of their ontological commitments toward the basic concepts and schemas of chemistry, has assisted us greatly here. It is molecular stereochemistry that now provides the main ontological basis of what nonetheless remains the genuinely biological science of genomics; and its schemas and models are by far the least questioned of all those employed to frame the detailed findings of the field. And these same chemical models merit a similar standing in the far less detailed accounts of genomics of which many nonspecialists now have need. We have tried to give them this standing in this book and they provide the default position for our own references to "what is really there."

1

By Way of Background

Inherited traits

Current science can be the subject of several kinds of history. The ideas, theories, and ways of thinking characteristic of contemporary scientific projects are often presented to wider audiences as a mythical history, frequently beginning with the work of some heroic "founding father." Even advanced scientific textbooks sometimes include these kinds of accounts, wherein students can be given a brief overview of those contributions to a field that remain important to its current practice. While mythical histories should on no account be read as descriptions of "what really happened," they do serve their actual purpose well, and indeed what follows in this first chapter is a scene-setting account somewhat of this sort, looking into the past, as conventional histories of genetics generally do, as far as the work of the nineteenth-century Austrian monk Gregor Mendel. There is, however, one particular limitation of mythical history that is inimical to our larger purpose. It can convey the false impression that a field emerged out of nowhere and then grew and developed independently of the context surrounding it, and it is important that we convey no such impression here.

We need first of all to remember that the kinds of ideas that initially informed genetics, the kinds of problems it addressed, and several of the methods and techniques with which it attempted to resolve them existed long before the science itself. Their origins lie lost in the mists of prehistory, whenever it was first noted of plants and animals, crops and herds perhaps, that later generations resembled the earlier generations that begat them. No doubt resemblances between individual parents and offspring would have attracted attention as well as resemblances between entire sets of creatures, and both kinds of resemblance would have stimulated particular interest where human beings were concerned. In any event, an active curiosity about the specific resemblances between creatures linked by descent and ancestry did emerge, and intensified over the centuries. Farmers reflected on their stock and how it might be developed better to serve their needs. And aristocrats reflected on themselves and the family portraits lining their stairs, all featuring the same prominent chin or nose perhaps, or the same worrying hints of the madness that "ran in the family." Inevitably, general questions arose about how, and why, and to what degree these ancestral resemblances existed. And out of such questions, together with the organization and professionalization of systematic empirical enquiry that had occurred in the preceding decades, there arose early in the twentieth century what we now recognize as the science of genetics.[1]

Classical genetics, the study of ancestry, must also be understood as a product of its own ancestry. Both the cultural resources for shaping its questions and the technical resources for addressing these questions already existed. Genetics inherited a rich set of resources for ordering nature and in particular a developed classification of the kinds of living things. Kinds of creatures are most commonly identified in adult form, and systematic accounts of these kinds and their visible characteristics, "phenotypical traits" as they came to be called, were available to geneticists. It was in the context of this inheritance that they initially came to ask how traits are "passed on" from adult to adult, just as it was through inherited methods, largely consisting of controlled selective breeding and the observation of its effects, that they initially sought to answer the question.

We should also recall another more technical but nonetheless basic ingredient in the cultural inheritance of geneticists, the cell theory. Geneticists took it for granted that complex organisms grow from a single fertilized egg cell, in the process known as *ontogeny*. The life history of an organism is a sequence of cell divisions, differentiations, and deaths.

---

1 For much valuable material on ideas on heredity that lay behind the emergence of genetics see Müller-Wille and Rheinberger (2006).

What does this imply about the nature of organisms? Is each organism produced anew by the process of ontogeny, or does an ever-present organism merely develop and change in the course of the process? The first formulation equates the organism with its final form. The second assumes that the organism is there throughout, present as whatever material object exists at any stage of development. This alternative recognizes organisms as constantly changing entities, not the outcomes of a line of development from A to B but entities traveling round an unending cycle of forms. On this view, organisms should be understood as life cycles. It is a view that will figure as prominently as the everyday view of living things in the course of what follows.

Human beings, and many other complex organisms, spend most of their lifetimes, and are hence most readily identified, as fully formed adult *phenotypes*. But there are organisms that spend relatively little time in this state and are far more often observed in other manifestations— many insects, for instance, spend months or years as eggs, larvae, and pupae before emerging for perhaps no more than a few hours in their full adult glory—and experience with creatures such as these might prompt the thought that organisms are better regarded as life cycles. Either way, however, the existence of traits and characteristics that recur in generation after generation, in adult forms linked by direct relations of descent, raises the question of how the traits are passed on. Of particular concern is the question of how distinctive traits are transmitted through, and hence presumably in some way represented within, the single cells that exist as the smallest and simplest stages in a life cycle. Of course we should not imagine that everything evident in the adult organism is also in the egg cell, since the egg cell is not the only material from which the subsequent adult is made and not the only thing that affects its manufacture. But if we notice the seeds of white flowers producing white flowers in their turn, and the seeds of purple flowers producing purple, all in flowers of the same species growing in the same environments, then it is difficult to avoid the conclusion that the phenotypical differences in color are related to invisible differences in the two sorts of seed from which the differently colored plants grow and develop.

All complex organisms are made of cells, and ontogeny is a process in which one cell becomes many. Single cells are tiny objects, ranging from about one micrometer (one millionth of a meter) for a typical bacterial cell, to about 20 micrometers for a typical animal (including human) cell, and up to 100 micrometers for some plant cells. We shall talk in later chapters about some important differences between the cells of plants and animals and those of bacteria,[2] but often we will talk as though cells were a fairly homogeneous kind. It is worth remembering that this is not really the case, though given our primary focus on genomes it will generally be safe to overlook the differences. One feature common to all cells is a membrane separating its contents from the outside. The fact that the explanations for many of the observed differences between complex organisms reside within the minute space enclosed by this membrane requires that we attempt to grasp imaginatively the world of the cell and appreciate its particular scale and dynamics. There are said to be ~10^{14} cells in an adult human body, all deriving from one initial cell, and the volume of an "average" cell is thus ~10^{14} of that of the entire body.[3] It takes a very high power of light microscope to reveal them as discrete objects and identify the nuclei that are their most obvious internal features. We might also reflect that at a leisurely walking speed of two miles an hour one could move back and forth across a cell ~10,000 times a second; very slow movements in our ordinary scheme of things can produce very rapid sequences of events in a cell. Later, when we talk of the things inside a cell as forming connected systems, a sense of the microscopic extent of the cell and the rapidity of sequences of events within it can help in understanding how profoundly the parts of the system are interconnected and how very rapidly they may affect each other.

Let us return to the process by which one cell becomes many. It is actually a repeated series of divisions of one cell into two. In the simplest case both the cell as a whole and the nucleus split into two to produce two cells of the same sort as the original. But complex creatures are made of cells of different sorts and occasional cell divisions involve cell differentiation and specialization as well, initiating lines of muscle cells perhaps, or skin cells, or blood cells. It is natural to assume that something in the cell itself changes to produce these different cells, visible in the whole organism as differences between muscle and skin tissue for example, but in any event cells do differentiate into a number of

2 Later we shall also mention archaea, which are grouped together with bacteria as *prokaryotes*. Organisms with (generally larger) cells including a nuclear membrane are called *eukaryotes*. In addition to plants and animals, the latter include fungi and numerous single-celled organisms classed as *protists*.

3 It is worth mentioning here that in addition to these cells derived from the initial human zygote, a human body contains (or carries around with it) some ten times this number of microbial cells, many of which are essential to its proper functioning. We shall return to this in chapter 4.

different kinds from which the bodies of complex creatures are made. Almost all of these are *somatic cells*, cells that function in ways that contribute to the viability of the creatures themselves, as muscle and skin cells clearly do in different ways, but one class of cells has a function of a different sort: *germ cells* or *gametes*, sperm and egg cells, are separated off from the routine functioning of the organism and reserved to initiate the next generation of organisms. And here matters are complicated in most complex organisms by the existence of sex. This may take many observable forms, from pollination in plants to the impregnation of females with sperm in animals, but it culminates in the fusion of two cells into one, the fertilized egg cell or *zygote*, the successive divisions of which eventually produce a fully formed adult of the next generation.

Crucial insights into inheritance involving sexual reproduction were provided by the work of Gregor Mendel on pea plants. Mendel took two varieties of pea that reliably passed on a specific trait and carefully crossed them, allowing the one to be pollinated solely by the other. Earlier we spoke of a variety of plant with consistently reproduced purple flowers and another of the same species with consistently reproduced white ones. Imagine Mendel crossing two varieties of pea of this sort and finding that he had produced a generation of entirely purple-flowered plants. Then imagine seeds being produced by allowing this generation of plants to pollinate themselves, and these pollinated seeds being planted in their turn. The result this time would have been a second generation of plants, mostly purple-flowered, but a significant proportion of which, in fact about a quarter, were white like the white "grandparent." Mendel obtained results of this sort in a number of crossing experiments. The results suggested that invisible factors were transmitting the traits via the seed. In the first generation each seed received a factor "for" white from one parent and "for" purple from the other. In the adult plants of that generation the flowers were all purple because the purple factor dominated over the white, even though the white factor remained present. But those plants did not breed true like their parents because unlike the parents they could pass either a purple or a white factor into their own pollinated seeds. Some of those seeds would receive white factors from both parents, and revert to being white-flowered in the absence of any purple factor to override the white. It is also easy to see that if plants with both a purple and a white factor are equally likely to pass on either, then about a quarter of the second generation of plants will receive two white factors and hence have white flowers. The factors cited here to account for Mendel's results subsequently become known as *genes*, and alternative versions of genes, such as those transmitting white or purple flower color were called *alleles*.

On this basis classical genetics established itself in the decade before the First World War, and many of the major achievements of that field over the next half-century arose from the systematic study of phenotypical differences in organisms as clues to how the Mendelian factors or genes were transmitted. The research involved took many forms. Phenotypical resemblances between generations were studied both in organisms well suited to incorporation in experimental systems and in those, including humans, where useful outcomes were envisaged. The latter kind of work, famously exemplified by a study of transmission of hemophilia through the royal houses of Europe, produced the family trees showing how factors or alleles are transmitted and expressed. These remain among the most widely familiar manifestations of Mendelian genetics and continue to be used to record the incidence of so-called single-gene disorders in human families today. Over the same period attempts were made to estimate the frequency of given alleles of a gene in entire populations, and resulted in the emergence of the influential subfield of population genetics. And a great deal of research was also done on mutations, changes in the factors or genes themselves, again inferred from changes in the associated visible traits of the organisms involved.

That the supposedly stable atoms of inheritance were actually prone to change was something anticipated from the very beginning of classical genetics, at least by those who took a naturalistic, evolutionary view of biological diversity. How else could the emergence of new forms be explained if the Mendelian factors did not change? A kaleidoscope of unchanging factors would not suffice to account for the variation and increasing complexity of life forms that was evident from natural history. The study of mutations proved a richly rewarding field of research, and the more so after methods of inducing mutations by use of radiation or chemical treatment were invented. With hindsight, we can identify these inventions as providing the first technology of intervention into the constitution of genes themselves, albeit a crude one with generally unpredictable results, and it has been extensively used, by plant breeders especially, over subsequent decades.

Classical genetics is often referred to as Mendelian genetics or Mendelism. And although this is in many ways misleading, the honor conferred on Mendel's work is amply justified both by its quality as empirical science and its theoretical significance. Mendel identified traits in his peas that were passed on stable and unchanged from generation to generation, and that could disappear, and then reappear as before, as though carried by ever-present invisible factors; and he analyzed his results in terms of stable invisible units combining and recombining with each other. This last aspect of his work was celebrated in later accounts of "Mendel's laws," familiar to generations of genetics undergraduates through constantly changing textbook presentations. These laws, however, were neither formulated by Mendel nor, as it turned out, true, and in referring to their role in the textbooks we are actually looking forward beyond even the famous "rediscovery" of Mendel's work in 1901–2 to the period following World War One, when "Mendelian" genetics flourished as a science in many locations, and most famously in the "fly rooms" where Thomas Hunt Morgan and his collaborators studied a fruit fly, *Drosophila melanogaster,* and its chromosomes.[4]

For a brief moment longer, however, let us continue to ignore chronology and look at one standard anachronistic formulation of Mendel's laws. Mendel's first law states that alleles of the same gene *segregate* during reproduction, and the second law states that alleles of different genes segregate and *recombine* independently of each other. The formulation dramatizes the great importance of the notion of an allele in the context of classical genetics. A gene, as we noted earlier, was the (hypothetical) factor responsible for the kind of trait studied by genetics; an allele was one of the different forms that the gene could take, inferred from the presence of a different form of the visible trait itself. In sexual reproduction individual organisms were imagined as carrying two genes or factors "for" any given trait and as able to pass on only one of the two to progeny, so that inheritance was thought of in terms of *pairs* of alleles and, in the course of sexual reproduction, the separation and recombination of the two alleles in the pairs. A crucial point to bear in mind is that only insofar as a gene had alternative alleles was it even a part of the subject matter of classical genetics. Classical genetics was a study of the transmission of *differences*; without different forms its methods left it with nothing to study. In practice it liked to focus its methods on binary differences involving two visible alternative forms and two alternative postulated alleles.

With the concept of an allele in hand we can now ask what Mendel's laws amounted to in the context of classical genetics. Think once more of the plant with the purple and white flowers. We assume that this difference traces to alleles P and W, which produce the purple and white phenotypes. If these are the only available alleles, any plant will either

---

4 For an excellent account of this classic work on *Drosophila*, see Kohler (1994).

have two P's, two W's, or one of each. We refer to these as the *geno-types* PP, WW, and PW. What the first law amounts to is the claim that in reproduction either allele is equally likely to be passed to a descendant. The alleles *segregate* in reproduction: a plant with genotype PW is equally likely to transmit the P and the W genotype. The second law states that all genes segregate independently of one another. So, schematically, imagine an organism with genotypes PW and CD. Whether one of the offspring receives the P or the W allele, the likelihood of its receiving the C rather than the D allele remains unchanged. This means that the offspring as a whole will find themselves with different combinations of parental alleles, as a result of their *recombination*. Sexual reproduction thus involves a radical reordering of the alleles in the organism, a shaking of a kaleidoscope as it were, in which the particles are individual genes.

We mentioned above that Mendel himself never formulated the "laws" we have just described and that they are false as general laws. Exceptions to the first law are somewhat esoteric and not immediately relevant here, but that Mendel's second law is false proved a source of extraordinary good fortune for classical genetics. To understand why, we need to return to the key theoretical conjecture of Mendelism, that visible traits were passed on by stable, discrete, invisible "factors," subsequently to be called genes or alleles. If this was indeed occurring, then the factors had to pass through a single cell, and the question arose of whether or not they could be identified with any part of the cellular material. Prominent among the candidate material objects here were the most substantial and easily observed components of the cell nucleus, the chromosomes, which appeared under the microscope in favorable conditions as elongated objects, constant in number in all the cells of any particular organism.[5]

To be precise, the number of chromosomes was a constant in all the somatic cells of the organism. In the germ cells or gametes, only half the chromosomes of somatic cells were apparent. Moreover, the chromosomes of somatic cells were apparently made up of like pairs, and those of a gamete consisted of one of each pair, so that each gamete carried only half the normal somatic cell chromosome complement prior to fu-

5 In fact, chromosomes were for a long time extremely difficult to observe, and especially so under their natural conditions. Even though their role in inheritance was recognized from the early 1900s and they were extensively studied, agreement on the number and size order of the chromosomes in humans was not reached until the 1970s.

sion. It seemed clear that during sexual reproduction the chromosomes segregated and recombined just as the Mendelian factors were thought to do, as stable independent units, which made it extremely tempting to locate the Mendelian factors in the chromosomes. But while there was a very large number of factors, or pairs of alleles, there was only a small number of chromosome pairs. If the alleles were indeed on the chromosomes, then the alleles "for" many different traits had to be carried by a single chromosome, and such alleles would presumably not be passed on independently but in conjunction. There would be *linkage* between some alleles but not all: whatever organism inherited a given chromosome would inherit all the alleles on it in a bundle. Mendel's second law would be false.

Let us now return to the work of Thomas Hunt Morgan. As so often happens in biology, one of Morgan's greatest achievements had proved to be his choice of an organism to work with and his organization of systematic work around it. The fruit fly *Drosophila melanogaster* was almost perfectly suited to the development of genetics. Although work in the fly rooms remained tedious and laborious, as so much highly successful science tends to be, and thousands of hours had to be spent examining flies one by one and sorting them into categories according to traits of interest, research now moved forward relatively rapidly. Most importantly, the fly could be bred easily; the life cycle was only a few weeks; it had a suitable suite of natural variations for study;[6] and its chromosomes were particularly suitable for microscopic examination. The large numbers of flies that could be bred and examined made possible statistically powerful analysis of the relations of traits between parents and offspring over several generations. The work in Morgan's fly rooms was far from being the first to show empirically that Mendel's second law was incorrect, but, as well as putting the matter beyond doubt, it was the work that most successfully exploited the deviations from the law.

In brief, empirical study of the transmission of the traits associated with them indicated that some alleles segregated and recombined independently but others did not. And this made it a reasonable conjecture, subsequently to solidify into accepted knowledge, that alleles were located in groups on the different chromosomes: the shaking of the

---

6 Some of these variations were more Mendelian in their inheritance patterns than others, and Kohler (1994) nicely describes how important it was to identify and sustain pure lines of genetically well-behaved flies and focus attention on them.

kaleidoscope of Mendelian factors in sexual reproduction evidently only reordered the chromosomes, and hence produced a less radical reordering of alleles than the second law predicted. But the tendency of the alleles of different genes sometimes to be passed on as if they were linked together afforded a wonderful opportunity to map genes onto chromosomes. By careful observation of different visible traits, and whether they were inherited independently or linked together, it was possible to identify clusters of genes (factors/alleles) and attribute each cluster to a specific chromosome.

Empirical work on linkage rapidly produced impressive results, but it also revealed that existing conjectures about the micro-realm were still far too simple. And again the findings that exposed their limitations were also a source of advance and opportunity. The linked phenotypical traits that confounded the second law were not *completely* linked. Being located on the same chromosome, on the same piece of material, did not always ensure that two genes (or alleles) were passed on together. The reason was to prove profoundly interesting. Chromosomes did not pass through the reproductive process unchanged. In the course of the cell division that produced gametes, the process known as *meiosis*, the chromosomes exchanged material: a degree of *crossing-over* occurred between the chromosomes in each pair, so that whichever chromosome was eventually constituted into the zygote, as the inheritance of some subsequent organism, carried material from both itself and its opposite number. Sexual reproduction did not create genetic diversity merely by shaking up chromosomes after all: far more diversity than that was engendered, if not as much as shaking a kaleidoscope of fully separate genes might have provided in theory.

Far more is involved in the process of meiosis, which produces gametes, than in that of ordinary cell division, *mitosis*, described earlier; indeed the sequence of events involved in the former is so "unlikely" that its existence in a vast range of sexually reproducing creatures is accepted as a sign of its status as a major evolutionary development. Crossing over, the breaking and recombining of the paired chromosomes, is perhaps the most remarkable of all these events and a natural analogue of what is now called recombinant-DNA technology, but what is of greater interest here is the way it was first exploited as a research tool. If a break is made in a physical object, those parts of it closest to each other are the least likely to be separated. Imagine a length of string on which are a series of knots: a random cut of the string is less likely to separate adjacent knots than others more distant from each other. Analogously, visible traits that are invariably inherited together will correspond to genes close together on a chromosome and traits that occasionally separate will correspond to genes farther apart. By applying this rule again and again it was possible to map the order of genes on chromosomes. Starting from the work of Morgan's collaborator Alfred Sturtevant, who first plotted the relative positions of six *Drosophila* genes in 1913, this led to extremely detailed genetic maps—maps of entire genomes, as we might say today. These maps systematically correlated properties of the macroscopic organism with material features of cell nuclei 10^{-4} cm across, establishing a theoretical bridge of remarkable predictive reliability across six orders of magnitude. They represent an extraordinary achievement that had important consequences.

Almost all classical geneticists were content to make reference to factors or even "genes" as markers of visible phenotypical traits, markers that could help to predict future recurrences of the traits should they disappear for a time, or "skip a generation" as was sometimes said. But while some researchers preferred to remain agnostic about them and to use them simply as tools, others with materialist inclinations speculated that they existed as tiny discrete particles of matter. And with good reason, this latter view became increasingly widespread over time. It was hard to understand mutation, and especially its induction by radiation and chemical reagents, other than as a transformation of matter. It was no less hard to understand the way that genes were localized on chromosomes in any terms other than those of a materialist beads-on-a-string or links-of-a-chain model.

The main chemical constituents of chromosomes were protein and deoxyribonucleic acid (DNA), and the likelihood was that one of these two polymeric molecules, or perhaps both in combination, constituted "the genetic material." The story of how DNA came to be identified as the sought-for substance need not be retold here, but one of the many findings that clinched the identification is worth recalling. Viruses, the simplest known living things, so simple that they are sometimes said not to be living things at all, are largely made up of protein and DNA. Bacteriophages are viruses that infect bacteria. The single-celled bacteria are invaded by the far smaller viruses and, in many cases, their cell membranes eventually burst, releasing large numbers of new phages to infect yet more bacteria. Crucially, however, it was found that only the DNA of the phage passed into the bacteria, and its constituent protein remained outside the cellular membrane. Whatever went on in the bacterial cell subsequently was evidently brought about by the insertion of

DNA, not protein. It was phage DNA, not phage protein, that effected the reproduction of the bacteriophage.[7]

Phages inject DNA into bacteria, which then produce new phages. This provides strong support for the conjecture that DNA is the genetic material, but it is not the only reason that we have singled out this particular finding. It is the vivid way it illustrates what is involved in *being* the genetic material that has prompted us to highlight it. Phage DNA does not produce new phages. It is not in itself a sufficient cause of phage reproduction or an adequate explanation of the characteristics that phages manifest. Indeed, in the phage itself DNA does not do much at all: viruses may actually be crystallized and stored indefinitely in that condition. New phages are assembled by a system of production most of which is that of the infected bacterium. The explanation of phage reproduction needs to refer both to the constituents of the bacterial cell and to the DNA of the phage. Even an explanation of the particular features of the phages that are passed on in their reproduction needs to refer to both these things. All by itself, the DNA accounts for nothing at all. The causal processes that go on in organisms are intelligible only as ongoing processes in systems in which lots of components have roles. The DNA inactive in the phage beautifully symbolizes its lack of causal power in isolation from a larger reproductive system, which in this case resides in a bacterium external to the organism reproduced, whereas in more familiar cases it lies alongside the DNA in the cell of the reproducing organism. But if DNA plays only one role among many in the process of reproduction, we may even wonder why it should be singled out for identification as the genetic material.

The crucial point is that here as elsewhere DNA is the key to the understanding of difference and variability in what is reproduced, and it is common to distinguish causes not just as factors involved in the production of effects, but more specifically as those factors that explain differences in the effects.[8] The difference in the internal operation of the bacteria prior to and following infection by phage may legitimately be said to be caused, or brought about, by the phage DNA that is injected into it. Similarly, the differences in phenotype correlated with Mende-

7 The crucial experiments are reported in Avery et al. (1944) and Hershey and Chase (1952).

8 Alternatively, causes may be those special factors we cite to explain, given what we take to be normal or "background" conditions, why an effect occurred at all. But this can be understood as a special instance of causes as difference makers where the effect is itself the difference.

lian factors by classical geneticists also correlate with differences in DNA and may often be said to be caused by them or explained by them. Other contributory causes exist in these instances of course, but in the framework of the kind of causal discourse just described, the phage DNA is the change in the system that brings about the changes in which we are interested and is correctly referred to as the cause of those changes. Contributory causally implicated factors that are common to bacteria in which these phenomena do not occur are normally referred to not as causes but as background conditions, necessary for these changes to occur. References to DNA as cause, then, may be said to solve the key problem of classical genetics, that of explaining phenotypical differences, even though it is not "the full explanation" of traits or even of differences in them. The case of phage reproduction symbolizes this important point beautifully. The bacterium does one thing prior to the injection of phage DNA and another thing afterwards. The DNA injection makes the difference.[9]

Inherited molecules

Once they were recognized as being more than mere markers, genes and alleles became typical examples of the kinds of invisible theoretical entity central to many kinds of scientific investigation, not least in the biological sciences. They were assigned functions, and their unobservable

9 The importance of making a difference is familiar in philosophical analyses of causation. We say the lighted match caused the conflagration in the barn, even though we are well aware that many other factors, such as the presence of oxygen, are also required for the barn to burn. The match is the factor that explains the difference between the burning barn and its more fortunate neighbor: both contain oxygen, but only one contains a lighted match. Where a high-temperature process is carried on in an oxygen-free environment, by contrast, it is the accidental infiltration of oxygen to which we attribute any disaster. We are interested in causes as providing explanations, predictions, and capacities to manipulate situations. In all these cases we are looking for factors that make a difference. As we will consider later, there are other kinds of scientific investigation in which we are more generally interested in how things work, and in such contexts the importance of difference makers is less central. The classic discussion of the distinction between the total set of causal conditions required for an effect and pragmatic grounds for distinguishing particular members of this set as "the" cause is Mackie (1974). For an emphasis on the importance of causes as possible bases for manipulation of situations, a point increasingly relevant to our interest in DNA, see Woodward (2003).

functioning was cited to account for observable features of the organisms in which they functioned. When genes were subsequently associated with a chemical substance, this functional orientation of biology confronted the more structural orientation of chemistry, with fertile consequences. Objects identified in terms of what they *did* become at the same time objects identified in terms of what they *were*, and a marriage of function and constitution was achieved in new forms of scientific understanding. Genes were now material objects defined and identified in terms of function; they were, in fact, to be the carriers or transmitters of macroscopic traits.

The material of which genes were made had to be of a kind that would account for the doing of the things that genes did: its chemical structure had to give it the capacity to function as genes functioned. On this basis the requirements of the genetic material could be, and were, specified. The material had to be complex and capable of a very high degree of variation about its basic form since one of its functions was to transmit the quite extraordinary range of variations in visible pattern and form manifest in the macroscopic realm of living things. It had to be stable if that variation were to be transmitted with the accuracy and reliability attested to by observation, and yet some small degree of instability was required to account for the mutations both encountered as empirical phenomena and required if biology were to account for how living things had evolved over time. And finally the structure of the material had to be such that it could rapidly and accurately be replicated, given that copies of the genes even of a single individual organism passed into countless billions of sperm cells, each itself capable of transmitting the full range of the organism's phenotypical traits. For some time, awareness of these requirements inclined scientists to identify the genetic material in the chromosomes as their constituent proteins, since proteins, the basic "building blocks" from which the bodies of complex creatures are made, were known to be sufficiently stable and highly variable molecules. But eventually it became clear that DNA possessed these attributes as well, and as more and more was learned of its chemical structure so more was understood of how it was capable of functioning, in the cellular context, in the ways required of genetic material.

It is customary to identify the culminating, revelatory event here as the discovery of the stereochemical structure of DNA by Watson and Crick in 1953, and indeed their famous helical model of DNA not only shows how the molecule can incorporate systematic variation but it also suggests a possible mode of replication of the molecule that preserves that variation intact. DNA was already known to be a polymeric molecule of the requisite stability and often of a quite extraordinary length, giving it the capacity to embody all the complexity and variation any geneticist could desire. It was also known to contain variable proportions of four components, deriving from adenine, cytosine, guanine, and thymine, chemical substances known as organic bases and themselves of complex molecular structure. Acceptance of the Watson-Crick model enriched existing knowledge in a number of ways, but crucially it identified the organic bases as strung out along a stable linear chain in ordered sequences, with two such chains linked together in parallel through a weak form of chemical bonding that joined thymine always to adenine and vice-versa, and cytosine always to guanine and vice-versa. The linear sequences of bases were highly complex and variable as required. And the complementary base-pairing that bound two strands of polymer together into the double helix of the model indicated how the material could replicate in a way that preserved the precise sequence of bases in their existing order: if the complementary chains were to separate, each single chain could provide the template for a new complementary strand, generating a new double helix. The structure of protein, on the other hand, gave no indication of how it could bring about its own replication, and the conviction grew that DNA could provide an understanding of protein replication as well as its own replication, and that the variability in the phenotypes made from proteins had to be passed on as variability embodied in DNA.

Much of the subsequent work on the functioning of DNA has employed an informatic model, in which the variable chemical structures of the four organic bases have been represented by their initial letters A, C, G, and T and the base sequences have been treated as texts written along a material substrate rather as messages may run along the tape of a ticker-tape machine. The two key functions of DNA are now commonly described in terms of this model. First, it must replicate and in this process the letters of the text are "read" singly, rather as instructions are read, with A calling for T and vice-versa, and C calling for G and vice-versa, so that the sequence on a length of DNA may be imagined as the "instructions" for the manufacture of its complement. The complement may then in turn serve as the instructions for the production of its own complement: this, of course, is the sequence of the original length of DNA, which is thereby replicated.

Second, as the distinctive forms of organisms are constructed from a great number of different kinds of protein molecules, which are long linear polymers like DNA itself, it was expected, and soon confirmed, that DNA made a difference to the macro-constitution of organisms largely through its having a role in the synthesis of different proteins. Here the text of the DNA is "read" not as letters but as words of three letters. Proteins, of which many thousands are now known, are all made from a very small number (around twenty in the case of primates) of constituent amino acids, and a lexicon was painstakingly constructed, the legendary "genetic code," in which a three-letter word or *codon* stands for a given amino acid. Read as a sequence of three-letter codons, a sequence of bases could be understood as the specification for a protein made of a particular sequence of amino acids. Here, the reading is a more complex affair than that associated with replication. First the DNA is "transcribed" into shorter strands of RNA, ribonucleic acid, a polymeric molecule chemically very similar to DNA but with uracil, U, replacing thymine, T, in the four-letter text it carries.[10] The RNA "text message" is then "translated" into the protein itself. Despite these complications, the DNA sequence can be said to specify the amino acid sequence of the protein. And we have the amazing phenomenon of a natural ordering that read in one way specifies its own replication, and read in another specifies the ordering of another kind of molecule altogether.

We are now close to the limits of topics that can reasonably be counted as background. But before concluding this preliminary discussion we should say something of genomes, the objects that, according to its title, our book is about. It will not be until chapter 3 that we discuss genomes and how they are best understood in any detail, but something provisional is required in the meantime and can be provided via an analogy with genes. Once the genetic material was identified as DNA, a gene was naturally thought of as a bit of that DNA. To sequence an appropriate bit of DNA would identify the specifications for a particular protein. Suppose, however, that we could isolate and sequence all the DNA in the chromosomes of an organism. Then we would have the specifications for all the proteins in that organism. We would have the sequence of the *genome* of the whole organism, and within that sequence, somewhere, would be the sequences of all the genes. And thus, it was supposed, the human genome, defined as the entirety of the DNA in a human cell, would contain all the sequences relevant to producing a human, the *Drosophila* genome would have all those relevant to producing a fruit fly, and so forth. We shall soon encounter a number of problems and complications associated with this way of looking at things, but it will serve for the moment as we move on to discuss genetics and genomics as practices and activities. And indeed for all its imperfections what

10 We shall say much more later about RNA.

we have just sketched remains perhaps the most common way of understanding what genomes are.

Practices and techniques

Genetics and genomics, like all sciences, are *done*, and the scientists who carry and transmit their knowledge are researchers as much as or more than teachers. But when we try to take this into account by switching perspective and considering sciences primarily as forms of activity rather than as bodies of knowledge, a profound shift of awareness occurs that can be difficult to handle. Even our understanding of what knowledge itself consists in and what it is knowledge of is liable to change. In the everyday world it is perfectly sensible to think of people relating to familiar material objects with stable properties that we know of and experience as properties of the independent objects themselves. But scientific research is less concerned with known and familiar objects than with bringing what is unknown and unfamiliar into the realm of the known, and the objects about which it is most curious are not those of the everyday world, or even those of natural history, but rather theoretical objects that are invisible and intangible. Both the existence of such objects and their state and condition at any given time have to be inferred indirectly. This is why so much of the time of those who research on genes and genomes, which clearly feature among these objects, is spent on developing methods of following things with markers, methods of testing for things by seeking signs of their functioning, methods of separating materials supposedly inhabited by different kinds of things, methods of exposing signs of things in population data by use of statistics and other mathematical methods, and so on. Indeed, to work in this way, or to study closely those who work in this way, can prompt the thought that the methods by which these objects are detected and studied are more important than the objects that they apparently detect; and that references to the latter may even be no more than ways of describing the esoteric practices of scientists. The history of science, in turn, may then best be thought of as primarily a history of changing methods and practices oriented to the study of specifically scientific objects, or epistemic objects as Hans-Jörg Rheinberger (1997) has called them. And in defense of this perspective it can be said that the methods and their application are there to refer to as visible entities, whereas the supposed objects are not, and—more tellingly still—that it is often the case where methods of research change that the invisible objects that scientists believe themselves to be studying change as well.

Historical and sociological studies of the biological sciences are increasingly seeking to do justice to the role of methods and practices and the artifacts they deploy (Rabinow, 1996; Rheinberger, 1997; M'charek, 2005). Studies of this kind offer fascinating accounts of what has actually gone on in research laboratories as well as of the sheer complexity of the processes and procedures through which a shared understanding of their invisible contents was fashioned. But while they convey a strong sense that this is indeed how scientific research ought to be studied and understood, they demand a mastery of detail that puts them beyond the scope of the discussion in this book. Even so, we want to highlight the epistemic importance of addressing genetics and genomics as methods and practices and not merely as descriptions of how things are. And we shall at times call attention to the methods by which knowledge has been produced in these sciences, almost invariably in order to highlight the need for caution when drawing inferences from it or generalizing on the basis of it.[11]

One important feature of genetics and genomics as practices is the use they have made of nonhuman species. Our knowledge of human genomes derives from the study of animals and plants, bacteria and viruses, as well as from the study of human beings. We cannot do things with and to humans as readily as nonhumans, and knowledge derives from manipulation as much as it does from observation. So researchers have allowed their knowledge of humans to be generated by the extension of what has been learned of other species, as well as from what has been ascertained directly. They have assumed the validity of the analogy between humans and other creatures, and have often also assumed that humans and living creatures generally can all be studied as different but related kinds of material things. There is no doubt that a materialistic and naturalistic approach informs the practice of these sciences, and that it amounts to their acceptance of a prior assumption about the nature of human beings.

Historically, human genetics, and hence genomics, has emerged from the study of the entire living world. People habitually manipulated the bodies of other creatures in the course of their economic activity, and research in genetics was able to systematize the existing manipulative techniques established in such practices. Early Mendelian genetics, for example, grew out of the existing practices of plant and animal breed-

11 Much of the notorious misuse of the knowledge of the old genetics has involved overgeneralization, and the same could easily occur with the knowledge of genomics, so the issues here are far from being trivial.

ers, which in turn were a part of a larger agricultural economy. Mendel's own work is worth recalling at this point. While undeniably highly creative and original, it relied heavily on an inheritance of skills and techniques developed by plant breeders. Even the material objects the breeders had produced were crucial resources for Mendel. His work was illuminating because it was done on pure lines of pea plants. The particular pea plants he used constituted, unbeknownst to him, a fortunately structured system from which he would derive results of rare pattern and simplicity. Much of the careful design and ordering of the empirical situation usually necessary to the production of simply patterned research results had been done for him already by plant breeders. And more generally, traditional Mendelian genetics was the analogue in the context of science of plant and animal breeding activities in the context of the economy. Breeders proceeded by crossing, selecting, eliminating, and selecting once more. Geneticists used the same procedures, and were sometimes able to suggest potentially profitable crossing and selecting strategies for breeders to apply.[12]

Manipulation of this sort was essential to genetics research; therefore, such research tended to focus on nonhuman organisms. Indeed, fears about the potential application to humans of techniques that have been so efficacious in the manipulation and economic exploitation of other creatures have formed the dark shadow of genetics and genomics since their inception. But the actual practice of these sciences has increasingly focused on just a very few kinds of living things, specifically selected for their amenability to manipulation rather than on account of their position in the natural order. Thus it was that *Drosophila* became the basis of a productive experimental system, to be followed by the fungus *Neurospora crassa*, the bacterium *Escherichia coli*, and many others, with attention moving along from one to the next partly as the result of technological advance altering the specifications of the "ideal research object" (Rheinberger, 1997). Findings generated in these experimental systems were extended by analogy from the organisms studied to other kinds, including humans. And much insight and understanding were produced thereby, at the cost of some occasional false trails on

12 To illustrate again the more nuanced historical analysis that is possible here, the reader might consult Müller-Wille (2005). Müller-Wille explores the way in which, in the context of an important breeding station, traditional breeding practices provided serious epistemological obstacles to the implementation of Mendelian experiments. On the other hand, he argues, these obstacles actually proved conducive to scientific progress.

which, it can be said with hindsight, the analogies involved were pushed too far.

In the early period of classical genetics the most fecund experimental system was that based on *Drosophila* and created by Thomas Hunt Morgan and his students and collaborators. What was learned from this system, with relatively few simple techniques and artifacts, superbly organized and skillfully and efficiently deployed, was little short of amazing. As we have already described, Mendelian theories, as formulated after 1902, were radically modified and elaborated through linkage studies based on *Drosophila*. It is salutary to reflect on what these studies involved in the way of activity, in the sorting and counting of thousands of adult fruit flies, not to mention the work needed in maintaining the fly rooms. The mapping of genes onto chromosomes was based on the breeding, sorting, counting, and killing of flies. The remarkable maps produced stand as a tribute to what can be achieved through laborious slog in mundane activities like fly sorting. But it is worth noting as well that since representations and theories, here exemplified by maps, are always the products of specific methods and techniques, they can be trusted and relied upon only to the extent that the methods and techniques are relied upon.[13]

Morgan's group was mapping the chromosomes of just one species and the reliability of its results proved to be so high that they were a major source of encouragement to the "materialization" of the gene that occurred in the early decades of genetics research. The results also generated more and more interest in the question of what the hereditary material might consist in. We have already indicated how work on this problem resulted in the identification of DNA as the relevant substance and in the production of the Watson-Crick model of its spatial molecular structure in 1953. The present relevance of this is that it led to the importation of a host of new practices and techniques into genetics and to the profound transformation of the field both as practice and as knowledge that subsequently took place.

Revisionists have quite rightly challenged accounts of the history of genetics and molecular biology that make a fetish of DNA and especially of its double-helical structure. But it remains the case that the double

---

13 Maps, of course, always have their strengths and weaknesses, related to how they are made. It is necessary to be aware of how they were made in order to use them properly and to avoid misusing them for purposes they will not reliably serve. All this makes maps a wonderful metaphor for scientific knowledge generally.

helix has proved a wonderfully potent condensed symbol, and it can serve in this role now, reminding us of the great range of disparate and hitherto dispersed and unconnected techniques and practices that were brought together to create it. Watson and Crick ranged across all the scientific work on DNA they could find, looking for clues to its structure. They used research from several different experimental systems in order to create their proposed structure, incorporating data not just from biology but from X-ray crystallography, and from chemistry, in which Chargaff's work on pyrimidine/purine[14] ratios furnished the vital clue to the existence of complementary base-pairing. The helical model was the visible outcome of the weaving together of all this research and more. It could serve accordingly as a badge of honor for practitioners in many different institutional settings, a memorial to the history in which they had all played a part, a sign of the present and future possibilities of cooperation within the strong division of technical labor in science, and a reassurance that science remained a connected whole despite the existence of that division of labor.

The double helix as a symbol is still more familiar and widespread today than it was in the 1950s and '60s, and it now reminds us of how the different traditions of practice that made it possible have combined to exploit and build upon it. For in the following years a major technical and procedural reorganization was to occur in biology that was also a social reorganization of its constituent fields. Effort and resources shifted more and more to the micro level: from the organism to the cell and from the cell to its constituents. Molecular biology and molecular genetics began their remarkable expansion, and the helix became an emblem of the molecular approach that continues its remarkable success today, notably in the guise of genomics. The old Mendelian genetics elicits amazement through what it achieved with a limited and slowly changing technology. Industry and organization, plus the astute choice of organisms and experimental systems, had made the very most of what could be absorbed via the eye alone or through the light microscope. That observations of phenotypical variation allowed chromosomes to be mapped at a resolution sometimes as fine as a few base pairs—as we would say today—is eloquent testimony to what industry and application can achieve. But

14 Purines (adenine and guanine) and pyrimidines (cytosine and thymine) are the bases that bind to one another across the complementary strands of the double helix. Chargaff found that the ratio of pyrimidines to purines in organic DNA was 1:1, which is an immediate consequence of the pairing structure discussed above.

the success of postwar genetics is testimony to what can be achieved through the transformation of technology itself, which increasingly allowed researchers to deal with cells as readily as with macroorganisms, and then to deal with molecules, "theoretical entities" though they were, in some ways as readily as they dealt with cells.

A whole range of new instruments and techniques became available to (molecular) geneticists in the 1960s and '70s. Electron microscopes, X-ray techniques, and a range of spectroscopic methods extended the threshold of observation so greatly that it became tempting to talk of molecules as observable entities, and irresistibly tempting to infer their characteristics from the signs and shadows of them that could now be observed. Various devices for tagging and labeling, particularly the use of radioactive isotopes, made it simpler to follow molecules around and discern the processes and events in which they were involved. Centrifuging and gel electrophoresis permitted the separation and sorting of molecules by size and shape. And powerful analytical techniques made it possible to infer the chemical composition of molecules, and in particular the sequences of the components of the major polymer molecules: DNA, RNA, and protein. The result has been a technology of staggering potency, allowing biological researchers to explore the realm of the cell nucleus at the molecular level almost as they please. And of course it is a constantly changing and growing technology that serves among other things as the means of its own expansion, prompting the thought that whatever at any given time lies beyond its reach is unlikely to remain so indefinitely.

The astonishing detail with which researchers can now investigate molecular structures and processes is not solely the consequence of imported techniques and instruments of the sort we have just listed. The technology of biology also exploits the remarkable characteristics of living things themselves, and their material constituents. Indeed, living things often *are* the technology of biology. Organisms, and even more their complex molecular constituents, are often remarkably specific in how they act upon each other: an organism may metabolize one substance but not another apparently almost identical one; a molecule may attach to one site on a cell surface but not another; within the cell an enzyme may function by acting only upon molecules with very highly specific structural characteristics, which can make it a potent and versatile tool for the researcher. Enzymes of this last sort have had a crucial role in facilitating the final technological advance that needs to be mentioned here, from observing, following, sorting, selecting, and analyzing organisms to intervening in their internal structures and modifying their functioning at the cellular and molecular level. By the use of *restriction enzymes*, first discovered functioning naturally in bacteria, researchers have been able to cut strings of DNA at sites defined by specific base sequences, and to manipulate the resulting fragments of DNA. This has been crucial for the development of recombinant DNA technology, which permits the movement of selected sequences of DNA from one genome to another or, as it is often described, the transfer of genes from one species to another.

The advent of recombinant DNA technology in the early 1970s has led to the production of many kinds of transgenic organisms, more recently known as genetically modified (GM) organisms, and to continuing ethical argument and controversy focused upon it and them. And the familiar implications of this technology entirely justify this deep concern. Just as what some regard as the given genetic basis of the nature of living things and the differences between them was being made accessible as a readable text, the power was developed to rewrite that text almost at will, a power capable both of dissolving and transcending that supposedly given basis. But there is more to be worried about here than the familiar implications of this potent technology. The powers it provides to intervene in the innermost structures of living things are likely to grow and develop in ways that are as yet hardly imaginable. What the consequences of this expansion will be, and indeed what the consequences of the routine use of such powers that already exist will be, it is impossible to say, but they will surely be profound, and anxiety is an entirely appropriate response in the face of them.

Another key component of current technology is of course the sequencing machines, now automated and capable of coping with immense throughputs, that "read-off," rapidly and reliably, the four-letter sequences carried by strands of DNA. By the 1990s entire genomes were being sequenced, and the new fields of genomics and bioinformatics were defining themselves through their focus upon these objects and the messages they carried. Within a few years the sequences were being directly loaded onto computers and the Internet, to be stored and processed with increasing ease by an ever more powerful information technology, and a subgroup of researchers emerged for whom genomes were series of letters existing only in silico, and genomics was their interrogation and comparison by use of sophisticated computer programs. The number of sequenced genomes and their size increased rapidly over this period, through work facilitated by continuing technological innovation and what quickly became a sea change in levels of funding. The potential utility of plant genomics and microbial genomics had become abundantly clear by now and they were attracting substantial levels of support; and an investment vast beyond all precedent was made to permit the sequencing of the entire human genome. What prompted the latter is still far from clear, but the question is worth dwelling on as it raises intriguing issues about the nature of scientific activity and the relationship between science and technology.

The sequencing of the human genome was a project at the very limits of feasibility, at least when the work first began, and one where immediate applications for findings were not obvious, so it is natural to ask why the truly enormous resources it required were made available. Of course, any project of such size and scale is likely to be supported, and opposed, for many different reasons. There will usually be many different and even conflicting understandings of what its objectives are, and what kinds of activity are involved in it. Certainly, the Human Genome Project was very widely perceived as a scientific venture, indeed as the most spectacular and ambitious project ever undertaken by the biological sciences. But it was also widely perceived and criticized as of questionable scientific value, and at the time of its inception there was a significant weight of scientific authority in the anti camp as well as in the pro. At this time the valuable results of such a project, both in terms of scientific interest and of likely utility, were largely thought to consist in the sequences of the "coding DNA" that would be produced and the "genes" that would thereby be identified. And some critics suggested that a more efficient and inexpensive way of obtaining these would be to focus not on the DNA of the genome but on transcribed RNA, since in the human genome only a very small proportion of the DNA codes for protein.[15]

In the event, of course, the sequencing proceeded far more rapidly than first anticipated, with the task being greatly facilitated by the technological advances that occurred in the course of it. As capillary sequencers replaced slab sequencers, IT became increasingly powerful in rough accord with Moore's law,[16] and automated machine rooms were

15 What was then a perfectly plausible argument now seems much less so, given that far more of human genomic DNA is transcribed than was then thought and that the functions of the vast amount of transcribed RNA is now recognized as a research topic of major importance. We shall make brief mention of this in the next chapter.

16 Moore's law is the empirical observation, attributed to Gordon E. Moore, cofounder of Intel, that the power of computing technology doubles about every eighteen months.

established in a triumph of organization and routinization surpassing even the fly rooms of old, the project positively flew along, with output eventually being streamed directly onto the Internet. Even so, the questions remain of why the entire genome was sequenced, and why the financial support to produce so much material of no clear and proven interest at the time, scientific or utilitarian, was forthcoming. And it is worth entertaining the thought here that major sources of this support regarded the project neither as pure science, nor as science and technology applied to achieve clear utilitarian goals, but rather as pure technology, technology for its own sake, even if that sounds like a contradiction in terms.[17]

As an exercise in pure technology, the genome project has important similarities with, for example, the US space program. Here again things were found out about the physical world, and here again many of the things found out at great expense were not reckoned as of major scientific significance. But both projects were conspicuous displays and celebrations of technical capability, and one of the most clear and evident results of the support they have received has been the expansion of the infrastructure that underpins that capability, and the further refinement and improvement of the skills and artifacts that constitute it. In both cases, an enlarged workforce with relevant training and abilities has been established and the production of instruments and devices has been stimulated; organizations and hierarchies familiar with directing and controlling the relevant technical systems have been brought into being. In short, an immense, generalized capacity for technical action has been created, available for use in the future even if no immediately profitable use is apparent. An investment in capacities has been made, which would have been impossible, or rather illegal, if directed toward any particular product or the benefit of any particular company. Whole economies have been offered vast sociotechnical systems with which to play. Naturally, there will be an assumption that these playthings are

---

17 Sulston and Ferry (2002) describe how their sources of support insistently view the genome project in utilitarian terms and how the project has been funded as if its sole purpose and rationale has been technological development. They go on to say, however, that their research ought not to be treated in this way and that they themselves regard it as a part of science, involving, as science generally involves, a wish to satisfy curiosity about what the world is actually like. But these different ways of regarding the HGP need not be regarded as mutually exclusive, and indeed different coexisting conceptions of the role and purpose of scientific projects, and indeed of "science" in general, are the norm historically.

likely to prove useful at some point, but the actual anticipated utilities need not be specifically identified.

Whether or not support for the genome project was significantly motivated in this way, it has certainly had results of the sort just described. Genomics now exists as an impressive body of trained scientific and technical person power, equipped with a formidable array of instruments and artifacts and in particular an enormous sequencing capacity. And of course all this has been just a part of a much larger process of capacity building. Techniques to detect, harvest, amplify, and manipulate DNA have proliferated on a much broader front, as have methods of monitoring its functioning in vivo. Recombinant DNA technology has become ever more sophisticated, and capable of a wider range of applications as more and more sequence data have become available and genomes have become better understood. Thus, we are confronted with a transformation of powers as well as knowledge. Classical genetics allowed only very limited forms of manipulation and experimentation, little more than those of plant and animal breeding. And insofar as humans were concerned, scarcely any manipulation was possible at all: in societies of free human beings, human genetics could reasonably aspire only to observe and predict, and perhaps occasionally give warnings to unknowing carriers of genetic diseases. In contrast, genomics and the new molecular genetics go hand in hand with advanced technologies permitting detailed modification of the internal structure of the organism and its germ line. The transition has been from a science largely limited to observation, selection, and prediction to a science with an as yet unknown potential for manipulation and control. Those who learn the knowledge of these sciences will increasingly find themselves learning how to intervene: how to diminish or enhance, selectively destroy or selectively benefit, impoverish or enrich, whether or not they actually do any of these things.

No doubt the existence of this new power will have profound effects both on the wider society and on the future course of genomics itself though, as we have stressed, we have as yet little idea how this power will be exercised. We have new powers and capabilities, not a menu of products and applications. There are critics who attack the new science for this very reason, distrusting our ability to cope with such powers and lamenting how they are foisted upon us without prior democratic discussion. And the rationale for this criticism is not unsound. There is reason to fear the extension of our technical powers and capacities, not the least being that we are ferocious creatures perfectly capable of wielding them against each other to our mutual harm. Generalized powers to control the physical world are ipso facto powers to control each other, and this is particularly obvious where the powers of genomics as technology are concerned. But these are topics appropriate for discussion later in this book, not at its outset; they are among the possible consequences of the changes it describes, not part of the background needed to make its story intelligible.

2

Genes, Genomes, and Molecular Genetics

Genes and DNA

Now that the ground has been prepared, the discussion in this chapter can focus more specifically on the move away from classical genetics and on the distinctive characteristics of the molecular genetics and genomics that emerged from it. Classical genetics was concerned above all with patterns of phenotypical differences in successive generations of organisms. It postulated the existence of independent "factors" or "genes," passed on from generation to generation, in order to account for the patterns. Gradually, most researchers became confident that these factors were real material entities, and this strongly suggested that the different factors had to be more than mere markers or correlates of different traits: the factors surely had to have some role in causing the traits to come into existence. What that role was, however, remained an open question. The process of *ontogeny*, in which pollinated seeds or fertilized eggs grew and differentiated into adult organisms with different traits, while well described at the macro level, remained beyond observation and understanding at the micro level, where the factors were assumed to operate.

Accordingly, once it became generally accepted that DNA was the hereditary material, there was a strong and immediate incentive to investigate what it was that DNA actually did. Researchers hastened to study its biological functioning with the outcomes we described in the previous chapter, and it became increasingly recognized that the role of DNA both in ontogeny and in the mundane, everyday operations of the cell was in the first instance bound up with the production of proteins. A great deal was learned about what was involved here by molecular geneticists working mainly on very simple organisms, and later work on the sequenced genomes of both simple and highly complex organisms was able to link the functioning of their genomic DNA to its chemical/molecular structure. This was work that, rather than merely adding to and filling out existing knowledge, revealed the need for a number of major transformations in the basic framework of that knowledge, some of which we have already described. But we have yet to describe the particularly striking developments through which research that initially focused on DNA as the material substrate of genes has radically transformed our ways of conceptualizing genes and even led to the suggestion that there are no such things.

The shifts in the foci and interests of researchers we have just described are clearly apparent in textbooks of genomics and molecular genetics, and notably in how they define "gene." In one important text, a gene is defined as "a DNA segment . . . coding for an RNA and/or polypeptide molecule" (Brown, 2002, p. 524).[1] And in another it is "a DNA sequence that contributes to the phenotype of an organism in a sequence dependent way" (Sudbery, 2002, p. 26). In general, in such texts, whenever a gene is taken to be a real material thing, it is defined as a specific length of DNA with a specific biological function. Interestingly, however, that specific function is no longer the traditional one of serving as a unit of inheritance. Nor is this mere carelessness, indicative only of the lack of importance of formal definitions in the actual practice of science;[2] rather, we see here a symbol of a pervasive shift that has occurred in that practice and its goals.

In classical genetics, genes were units of inheritance, but what was "inheritance"? The notion was very close to that found in everyday discourse surrounding the institution of the family and the institution of property, and indeed the scientific notion surely derived from the every-

---

1 Polypeptides are the long molecular chains of amino acids from which proteins are made, but the difference between polypeptides and proteins will largely be ignored here and the custom of speaking of DNA as coding for proteins, imprecise though it is, will be followed.

2 For relevant further discussion see Waters (2004).

Hi Steph

I have a favor to ask — I'm looking for a reviewer to scan one of my thesis chapters. It would be a really brief review — ½ hour or less of your time.

Let me know if you are up for it —

Muchos Gracias, Kiza

day one. Genes were inherited very much as facial features or person-
alities were taken to be inherited in families, or bequests and rights to
property. It was this frame that initially structured curiosity in genetics,
and not the other frame widely encountered in biology, in which the notion of a life cycle is crucial. But the life cycle frame has subsequently
become predominant as the transition to the new genetics has occurred.
Followed through life cycles, rather than generations, living things are
objects with continuity in space and time, which merely change in size
and form. A series of life cycles may be addressed as a single changing
object, rather than as a series of different objects between which something has to be transmitted in order to account for the resemblance between later members of the series and those that come before. From a
life cycle perspective, the different organisms that we identify as the ancestors of a given individual over successive generations are recurrent
patterns in a continuing cycle of changes, and what part of the cycle
will count as "the organism" should be considered no more than an
arbitrary decision. The alga *Chlamydomonas reinhardtii,* for example,
a widely used model organism, spends most of its time as a single haploid cell (a cell, that is, with just one set of chromosomes). These exist in
two distinct mating types, and under certain conditions cells of opposite
mating types fuse to form a diploid (two-chromosomed) object referred
to as a zygospore. This has a hard outer wall and remains dormant until
conditions become more favorable, when it usually undergoes meiosis
and releases four haploid cells (see http://www.chlamy.org/info.html).
Whereas we are mainly familiar with diploid organisms that form haploid cells as part of the reproductive process, in the case of C. *rhein-
hardtii,* the active part of the life cycle is haploid, and the diploid phase
is produced in circumstances of stress as a sexual form of reproduction. Which form should be considered as the organism? The answer,
of course, is "either," "both," or "it doesn't matter." The moral is that
what is fundamental is the entire life cycle, of which no part has a privileged status as what the organism *really* is.

    To operate within the life cycle frame is not to deny the standing of
DNA as the hereditary material so much as to leave it behind, along with
the older frame in which that description made sense. If living things are
life cycles, it is natural to ask not merely what DNA does during reproduction and development, but what it does all the time, and what role
it plays as a part of the continuously functioning cells that are parts of
larger organic systems. And this indeed is what is asked by molecular
geneticists, whose technology now has the potential to deliver detailed
answers to questions of this sort. Thus, we find the science of genetics

shifting its center of gravity away from the study of inheritance to cell chemistry, and away from the goal of predicting and accounting for difference toward that of understanding biochemical processes. Instead of being spoken of as independent atoms of hereditary material, genes, conceptualized as DNA, are now referred to as parts of chemical/molecular systems within the cell, and specifically as sources of particular proteins. And further emphasizing the divergence from the Mendelian tradition, most of the genes in such systems, far from being invoked to explain phenotypic differences, are identified as sources of proteins with crucial "housekeeping" functions, expressed in all tissues and all the time.[3] Moreover, much of the most interesting work on these systems has been on the functional interdependence of DNA and molecules of other kinds, and on how different molecules of the same or different kinds may function together and operate in ways whereby they both regulate and are regulated by each other.

One striking illustration of the change of perspective involved here is the increased interest now being taken in how many "copies" of a given gene or DNA sequence are to be found in the genomic DNA of an organism (Redon et al., 2006). It is hard to find a place for such a question within a framework in which a gene is simply a unit of inheritance: what would be the role of several copies of a gene for brown eyes? But if genes are recognized as having other functions, things can look quite different. For example, if the production of other kinds of molecule is a function of DNA, there is an immediate rationale for the existence of multiple copies of genes. If one DNA strand has a maximum output, the need for a still greater output may imply a need for more copies of the relevant DNA. The existence of whole arrays of similar genes in the many mitochondrial genomes of a single cell has long been understood in this kind of way. Mitochondria, found in the extranuclear cytoplasm of cells, possess their own DNA. They function as energy producers, and the demand for large supplies of energy is taken to account for the large numbers of mitochondrial genomes per cell. But interest in numbers of copies has intensified and extended much more widely than this, and continues to do so as views on the functions of genes/DNA have

3 Of course this shift in meaning has not happened smoothly or even completely. As Moss (2003) persuasively argues, many of the problems with our understanding of genes derive from a pervasive tendency to conflate genes as markers of the inheritance of phenotypic traits (which he calls genes-P) and genes as sequences of nucleotides (genes-D).

changed and data from the genome projects has facilitated work on the topic. Currently, for example, there is a rapidly increasing interest in the production of untranslated RNA as one of the functions of nuclear DNA. RNA molecules have many cellular functions of their own but have short lifetimes and may need to be produced in quantity, so it is interesting to note that the DNA sequences that produce tRNAs, specialized RNAs that transfer amino acids to growing polypeptide chains, exist in hundreds of copies in human genomes.

Too little will be said in this book about research on whole sets or systems of molecules and not enough on how DNA functions along with other molecules in such systems. Fascinating and important though it may be, outside the context of professional science there is less interest in the internal "housekeeping" functions of DNA than in how variation in its sequences ends up being manifested externally as more or less striking qualitative differences in the phenotype. Genes have been redefined as protein producers because DNA functions as a protein producer, but only a few single proteins reveal their presence in an organism via striking phenotypical traits. Most of the genes that produce proteins are "tedious biochemical middle managers," as Matt Ridley puts it, doing a "boring chemical job" (1999, p. 52). Such jobs are boring because they are essential, and hence carried out not only in all of us, but also in creatures of many other kinds, and thus they lack any relevance either to what is distinctively human or to what makes one human differ from another. Of course, such genes may mutate and malfunction, and where the mutations are not lethal the resulting alleles may become interesting as "genes for" various inherited diseases and syndromes. But this takes us back to difference again, which continues to attract general attention in a way that the truly fascinating problem of how organisms normally function apparently does not.

So let us move on, and look at how research on DNA has raised the question of whether objects called genes actually exist at all.[4] All kinds of problems have arisen from conceptualizing a gene as a strip of DNA that "codes for" a protein, and we shall mention only a few of them

---

4 This question has raized increasing interest among philosophers of biology and philosophically minded biologists in recent years. See Moss (2003); the excellent collection of papers in Beurton et al. (2000); Fox Keller (2000); Dupré (2005). Empirical research on the diverse ways in which scientists in different areas of biology think of genes has been carried out by Karola Stotz, Paul Griffiths, and Rob Knight (Stotz, Griffiths, and Knight, 2004).

here.[5] First of all, the DNA that codes for a given protein is not a unique sequence, if only because the genetic code includes redundancies. We have to think of a number of different DNA sequences all of which may produce the relevant protein, all of which could constitute the given gene materially. Another way of putting this is to say that the normal allele of the gene may be realized by any of a large number of DNA sequences. At the same time, however, a large number of other sequences may produce functionally defective variants of the protein, so that if the lack of function gives rise to a specific disease or pathology, the "allele for" that disease can be realized by an even larger number of different DNA sequences, all departing slightly but crucially from the normal form. Thus, the contrast made in classical genetics between the normal and the abnormal allele in a pair may become a contrast between thousands of "normal" DNA sequences and an even larger number of "abnormal" ones.

Whatever a gene may be materially, it is not a *unique* four-letter sequence of DNA. But still more awkward is the finding that a gene is not a *continuous* DNA sequence either. In humans, the DNA that codes for a protein generally turns out to be separate pieces of DNA, close to each other on the genome, but with lengths of noncoding DNA separating them. The pieces of DNA with coding functions are called *exons*, and the lengths that separate them *introns*. While the whole sequence is transcribed into RNA, subsequent "editing" processes excise the introns, leaving only the exons to be further processed into proteins. Thus both exons and introns, not to mention various other DNA sequences "upstream" and "downstream" of the exon/intron set, are implicated in the actual production of the RNA that guides the manufacture of the protein. In a nutshell, studies of DNA as the material substrate of genes suggest that genes must be identified as complex, spatially discontinuous objects if they are to be identified as material objects at all, and this has inspired a fascinating program of research on their internal structure at the molecular level and its relationship to their functioning at that level. But while these studies have advanced our chemical and biological understanding of DNA, they have radically complicated the task of speci-

5 One particularly interesting problem that we shall not go into is that raised by alternative reading frames. Transcribed sequences of DNA, including coding sequences, may overlap; sequences may be read in either direction (so-called "sense" and "anti-sense" readings); and both of these phenomena may occur together. These findings strongly reinforce the view that the genome has no natural and unique decomposition into its constituent genes.

fying just what genes are, to the extent that commentators can now cite example after example of the difficulties that confront "gene realism."[6] If genes are objects, then they are objects that vary enormously in their constitution, and they are composite rather than unitary objects—objects only in the way that the solar system is an object, or a forest is, or a cell culture. Nor is it clear what should be included and what excluded by the boundaries of these composite objects: is a gene an exon set, for example, or does it comprise the introns as well? Problems of boundary drawing of this last sort, however, are more formal and linguistic than practical, and a realistic attitude to genes has to confront far more difficult problems.

Consider again the exons that have just been identified as the coding sequences of a gene. Exons are generally transcribed into RNA. But the RNA pieces produced by a set of exons may be joined together in more than one way, a phenomenon known as *alternative splicing*, which allows the same set of exons to "code for" several different proteins, in some cases hundreds or even thousands of them.[7] Alternative splicing is now thought to occur in at least 70% of the coding regions in the human genome. Any particular exon in such a set may be not only a part of the "gene for" protein X, but also a part of the "gene for" protein Y, and protein Z: the exon manufactures a component that gets incorporated into all three proteins. Given this, it is impossible to treat these three genes as three independent material objects or as distinct and separate bodies of material. In other words, these genes cannot be regarded as objects at all in the traditional sense; if they are to be spoken of as objects at all, some fairly substantial semantic innovation has to be undertaken. Here is a truly serious problem, and it is not created by a rare and exotic phenomenon. When alternative splicing was first identified it was thought to be an oddity. But its importance, especially in eukaryotic genomes, is constantly being reassessed and it now seems clear that alternative splicing is the norm in human (and many other) genomes.

The same problem of the lack of independent existence of different putative genes can also be posed by considering introns, the DNA in the gene for a protein that does not code for it. Occasionally, an entire exon/intron set—a gene for protein Y let us say—may be found within an intron of the gene for protein X. The same piece of DNA may be a

6 See Fogle (2000), and also the references in note 4, above.

7 One gene in *Drosophila*, called Dscam, has over thirty-eight thousand splice variants.

part of the exon of one gene and of the intron of another. It may function in one way as part of a chemical system involving one array of DNA, and function in another way as part of a differently defined array of DNA. This last formulation is important because it reminds us once more that when pieces of DNA are referred to as genes they are being conceptualized functionally, for example as protein generators. When material objects are grouped and classified by function, there is no guarantee that those objects will appear just once in the whole classification system, since there is no evident reason why one piece of material should not have more than one function. Just as a ruler may function as both a straightedge and a measure, so a piece of DNA functioning as a part of gene X may also function as a part of gene Y, whether its function in the two cases is the same or different.[8]

Even now we have scarcely begun to explore the complications encountered when genes are presumed to be DNA sequences with gene-like functions. Thus, for DNA to function in a gene-like way it must not only encompass exons and introns but many other kinds of sequences as well: "enhancers" and "silencers," "promoters" and "terminators," "insulators" and "locus control regions" are just a few of the function-denoting terms assigned to the small noncoding DNA sequences that play a part in the activities of genomes. An "object" including all these components is a complex composite entity, lacking in spatial unity. And a gene may not only lack spatial unity; it may lack temporal unity as well. Consider the "gene for" Huntington's disease, one of the most well-known and devastating inherited genetic diseases, which causes severe damage to the nervous system in middle age. The disease-causing allele is initially inherited as a normally functioning set of DNA sequences, which codes for the functional form of a protein that the organism requires. It is unusual only in containing within one of its exons an atypically long "triplet-repeat," a repeated sequence of the letters CAG. The initial sequence does not impede functionally satisfactory transcription, but replication processes are error-prone in relation to it. Repeated faulty replication results in ever longer CAG repeats in copies of the gene, until eventually its transcription produces nonfunctional protein. In brief, the supposed gene changes over time; it becomes, perhaps we should say, a different gene, or a different allele of the same gene. Rather than a gene for

8 It is interesting to note here that proteins too may have more than one function. There is growing interest in so-called moonlighting proteins, and proteins have been described as having half a dozen or more quite distinct modes of action in different cellular contexts.

Huntington's, we could say there is a gene for a gene for a gene . . . for Huntington's. Hence the late onset of the disease.

If DNA is the material substrate of objects we refer to as genes, then these objects are very strange ones. They are not unitary objects, as we have now repeatedly emphasized, but, more salient still here, they are not separable, independent objects either. A single gene may plausibly be identified as DNA, say as an exon set. But an entire set of genes cannot be thought of as so many separate pieces of DNA each of which is the material substrate of one of the genes. There is no way that snipping a human genome into twenty thousand pieces will produce the twenty thousand genes it is now said to embody. To vary the metaphor, the set of genes in a genome is not to be compared to a bag of marbles. Take a marble from the bag and the rest will remain, but take a gene from the set and other genes could well prove to be missing as well. The DNA that by virtue of what it does is part of the gene for protein X may well be, by virtue of something else that it does, also part of the gene for protein Y. This makes talk of genes awkward and untidy in comparison with talk of the related DNA, which is at the same time the more accessible "theoretical entity." For this and other reasons, the existence of genes may now be thought questionable in ways that the existence of DNA is not.

With hindsight, it is worth asking what sustained for so long the belief in genes as distinct units, each transmitting one particular trait, and the later variant according to which there was a DNA sequence for every protein. Although traits, or more precisely variations in traits, apparently transmitted by single genes, were familiar in classical genetics, so were traits related to many genes or difficult to relate to genes at all. There was never any warrant for regarding the first kinds of trait as universal or even as the norm, or never until Mendelian model organisms were selected to define the norm in laboratory settings, as Robert Kohler has described (Kohler, 1994). And when "genes for traits" became "DNA for proteins," it seems a little strange that biological reflection did not take more seriously the possibility that more efficient ways of synthesizing proteins might also have evolved. One DNA sequence per protein is highly inefficient, and on the face of it evolution might have been expected occasionally to favor efficiency over inefficiency. To manufacture every model of the Ford Mondeo separately, making even the tires, hubcaps, and wing mirrors of each independently, is clearly less efficient than to make them as we do, in one integrated, highly organized system of production. And addressing living things as highly organized systems is second nature to biologists.

At any rate, we can now see that many genomes, including human genomes, do indeed frequently operate like efficient car manufacturers, getting components from different sources and fitting them together in different combinations. This allows them to produce many closely related proteins from a single exon set, in some cases simultaneously and in other cases at different stages of human development. With the hemoglobins in humans, for example, an alternative splicing system permits the production of one version prior to birth, a second at a later stage, and yet another in the mature organism. And more generally, the identification of more and more complexly organized, efficient forms of transcription and splicing is being used to account for how in humans around a million proteins may be produced from little more than twenty thousand so-called genes. Of course, the profound importance of the findings we have been discussing far transcends their purely semantic consequences. Even so, it is worth noting how the "genes" now being referred to in the light of them are not genes in any previously accepted sense. And more significantly, scientists themselves are now beginning to express an increasing dissatisfaction with their concept of "gene" in the light of these and related findings (Pearson, 2006). All this raises the question of what the long-term future of this long-established and still very widely used concept will turn out to be.

Paul Griffiths and Karola Stotz put it thus: "the older concept of a particulate gene . . . is unable to accommodate the genomic complexities that are being discovered on a daily basis . . . the so-called 'classical molecular gene' is simply not up to the job."[9] Nor is this purely their opinion. It is rather their considered judgment based on extensive research on how the term is currently used by different groups of scientists and in particular on the high levels of disagreement among them on how to describe problematic arrangements of DNA. Moreover, it is a judgment that is consistent with both definitions of the gene cited earlier (p.48). So far we have been guided by the first of these and have discussed the difficulties raised by regarding genes as protein-producing lengths of DNA. But things go no better if a gene is defined as "a length of DNA that contributes, via its sequence, to the phenotype of the organism." Where DNA "contributes to the phenotype" via the production of protein the problems already discussed remain. A "gene" in the firefly, for example, a set of DNA sequences, codes for a protein molecule that glows in the dark, and hence it is a

9 Griffiths and Stotz (lecture at Exeter University 2004); see also Stotz and Griffiths (2004); Griffiths and Stotz (2006); Stotz, Griffiths, and Knight (2004); and their website at http://representinggenes.org.

trait of the firefly itself that it glows in the dark: the relation between gene and phenotype is simply the relationship between DNA and protein writ large, the relationship already discussed. Where DNA "contributes to the phenotype" in other ways, different, albeit no less serious, difficulties arise.[10]

There is currently intense interest in that part of the genomic DNA that is transcribed but not translated into protein. The advent of new methods of sequencing and the increased use of so-called gene chip or microarray technologies have facilitated the study not just of gene expression but of RNA transcripts generally,[11] and revealed how in human and other mammalian genomes well over half of all the DNA may be transcribed, a proportion greatly exceeding that assignable to coding DNA. This has drawn attention to how many of the different kinds of RNA molecules in a cell do not themselves code for proteins. Although it has long been recognized that RNA plays a role in cell chemistry and is implicated in the processes that regulate protein production, the sheer quantity and variety of untranslated RNAs and the demonstrable functional relevance of at least some of them suggests that their importance and hence that of the DNA from which they are transcribed needs greater recognition. Here after all is DNA that "contributes via its sequence to the phenotype" and falls accordingly under one molecular definition of "gene." But if the custom of denying this status to the "noncoding" DNA that is involved were to be set aside, the floodgates would open on the whole enterprise of linking traits to specific DNA sequences conceptualized as molecular genes.

In summary, there is now much less of a case than once there was for regarding genes as real things, and scientists themselves seem increasingly to be recognizing this. Although they continue to make sense of phenomena by reference both to genes and to DNA, and indeed it may even be that genes are spoken of and referred to more than they have

---

10 Note, however, that this example, chosen for simplicity, is atypical in that the trait involved is the "trait" of a single protein molecule multiplied. Were it taken as the norm, it could give spurious reinforcement to an older notion of genes as entities that carry the traits they are "for" entire within themselves, a notion that can encourage facile and misconceived forms of genetic determinism. In fact, DNA sequences "contribute to" innumerable traits that individual protein molecules cannot exhibit and individual "genes" cannot determine.

11 Application of these technologies has also led to the identification of very highly delocalized protein-producing "genes," in the functioning of which even material from different chromosomes may be implicated, and in this way too they have made the notion of gene as object more problematic.

ever been, accounts referring to DNA increasingly carry the greater authority—or rather the greater *ontological authority*. Where there is pressure to decide one way or the other, the assumption that genes really exist is the one that is likely to be questioned. The assumption that DNA is really there, functioning as it is known to function, tends now to be accepted in practice as essential to the prosecution of research. But references to genes are beginning to be regarded as optional extras, still of very considerable pragmatic value and often irresistibly convenient to employ, but not vital to scientific understanding, and liable to mislead if understood literally as references to real objects.

A realist conception of genes as objects made of DNA, or perhaps we should say, since our concern is with scientific practice and not with metaphysics, a dogged determination consistently to treat genes as such objects, quickly leads to problems so awkward and recalcitrant that the temptation to abandon such a stance must be very strong. And it can only be strengthened further by the fact that there is little to lose by yielding to it, now that realist inclinations can so easily be redirected onto DNA itself. Once DNA is recognized as the material substrate of genes, ontological authority can be transferred to our models and representations of the material and any functions we may have wished to attribute to genes can be hung on those models and representations directly, with an elaborate functional terminology now used to refer to various lengths and sequences of DNA (and also increasingly to other molecules intimately associated with it).

Of course, molecular genetics and genomics are only parts of a larger field, and it may be that in other areas of genetics, where it is harder to imagine how research could proceed without talking of genes, there is a correspondingly stronger sense that they are really there. An obvious example is in those areas of medical genetics where the central concern is the patterns of inheritance of disease genes. From the point of view of molecular genetics and genomics, "the" cystic fibrosis gene, for example, can be any of over a thousand DNA sequences known to code for functionally defective variants of the implicated protein; it is hardly a well-defined object. But where interest is focused entirely on the hereditary relations between those who carry and/or suffer from this disease, it is irresistibly convenient to refer to the transmission of a gene; and it is usually harmless in that narrow frame to think of "the gene" as something real.

It is important to remember as well that genetics and genomics are but a very small part of the overall body of specialized technical knowledge and expertise established in society as a whole, and also that references to genes are now increasingly characteristic of the discourse through which the findings of these fields are communicated in simplified form to other specialists, and whether through them or otherwise to wider audiences. Such references are likely to be more authoritative ontologically among these audiences than in the context of laboratory work itself. It is indeed an intriguing thought that gene-based models of biological processes, for so long fecund researchers' models, are now increasingly to be found functioning as simplifications of those models, and that their ontological authority may be gaining increased recognition within the common culture even as it is less and less recognized in the practice of the laboratory.

DNA

Let us review the story so far. Classical genetics referred to genes as invisible markers of the transmission of phenotypical characteristics, and then increasingly accepted that these same genes were real objects somehow implicated in the processes that produced those characteristics. Until DNA was identified as "the hereditary material" and the remarkable investigative techniques of molecular genetics had been developed, neither the objects nor the processes could be studied directly. Subsequently, however, even as the new techniques allowed genetics to move to a new level of empirical understanding, questions were simultaneously raised about its key concept of the gene. In the light of what emerged, talk of genes became increasingly difficult to sustain as a contribution to a realistic mode of discourse, and existing notions of a particulate gene could be sustained only by the constant expenditure of disproportionate amounts of effort. Since talk of DNA and associated molecules was much more easily defensible in a realist frame, and since it was now possible to hang the functions previously attributed to genes directly onto DNA itself, there was little to gain from treating genes as real objects. And the consequent shift of ontological authority from gene-based theories to DNA-based theories and chemical/molecular models of its structure and function is an important part of what now marks the divide between genomics and molecular genetics on the one hand, and the traditional classical genetics that preceded them on the other.

To refer to DNA is to refer to molecules, and to understand it as material objects made up of other such objects, whether sugars and bases and so forth or the atoms that in turn make up these things. A genome, being DNA, is also well understood as a material object, an extremely long molecule somehow squeezed into the cell nucleus. But although this conception of DNA and the genomes made of it has become widely accepted, genomes have increasingly come to be understood in a narrower frame that without denying that they are DNA molecules addresses them almost entirely as carriers of information, embodied in the sequences of nucleotides, now rendered as a sequences of the four letters, A, G, C, and T, that run along them. This informatic model of DNA ignores most of its characteristics as a material entity and focuses attention on what, in complex genomes like those of humans, is only a very small part of it, the "coding DNA" involved in the production of proteins. It is telling testimony to the dominance of this narrowly conceived model that at the height of its influence the rest of the genomic DNA was widely referred to as "junk."[12]

Narrow and limited as it may be, there is an enormous amount that stands to the credit of the informatic approach. It has modeled "coding DNA" very successfully as a text written in three-letter words employing a four-letter alphabet. And many insights into how its structure and functioning are related have resulted from treating DNA as a text, and from the use of other models and metaphors characteristic of an informatic perspective. There are several of these. It has become common to identify some DNA sequences as *instructions*, and even to characterize genomes as the bearers of all the instructions needed to produce the phenotypes of organisms. Another, closely related, notion is that a *program* is embedded in the base sequences of DNA, one that *governs* cellular processes much as a computer program is sometimes said to govern computational processes. And in popular writing DNA is occasionally referred to as a *master molecule*,[13] a crude formulation that has the virtue of emphasizing something true generally of these metaphors: they are all anthropomorphic metaphors, which rely upon our familiarity with relations between people, frequently hierarchical relations between people, when describing the relations between things.

What should we make of the use of these informatic models? The first thing to say here is that far from its being inherently objectionable to make use of models and metaphors in science, it is necessary to do

12 Users of the term have pointed out, however, that junk and rubbish are not identical. The latter is thrown out while the former is stored in the attic.

13 Fox Keller (1994) both evaluates this metaphor and reflects on how to evaluate such metaphors.

so. Researchers have to employ them to make sense in their own minds of the systems they study, just as as they have to do so to convey some understanding of those systems to outside audiences. Why models and metaphors are necessary and how they function are vast topics, but suffice it to say here that without the coordination of perception and practice around patterns that already exist in other contexts, systematic research could not go forward. Indeed, this is no more than an instance of the more general point that without the models and metaphors current in our shared everyday culture, social coordination quite generally would not be possible, and social life in the sense that human beings know it could not exist. The deployment of models and metaphors relevant to our present concern is just the redescription of things and goings-on in one context in terms of familiar shared patterns established in other contexts. If blinkered vision is sometimes encouraged by shared models and metaphors, it is the price that has to be paid for the coordination they allow. The esoteric discourse of current molecular genetics research is replete with models, metaphors, and other narrative devices that function in this way, but so too are the most simple and uncontroversial forms of biological discourse. Even to treat living things as organisms, that is, as organized wholes in which the parts have functions in relation to the whole, is to presume and apply a model. Another important ordering model, as we have seen, is that which takes an organism to be a life cycle, only a snapshot of which exists at any given time. And yet another ordering is imposed on this cycle when it is cut and opened out into a line. It has long been the custom to make the necessary cut at the point of greatest simplicity—the moment of production of the zygote, or fertilized egg—and to employ a model of linear development from that point in time as the basis for narratives of the ontogeny of the organism. And as is usually the case, the selection and use of such a model are conditioned by human interests and existing patterns of human understanding.

Since models and metaphors have to be used, it is pointless to criticize their use as such. What should be said, though, of the use of specifically anthropomorphic models? DNA is a material already conceptualized in terms of the authoritative molecular models of structural chemistry and stereochemistry, the models that identify it as deoxyribonucleic acid in the first place. If an additional bioinformatic model had to be layered on top of this, should it not first have been stripped of any gratuitous and potentially misleading anthropomorphic features? This is certainly the view of some critics. They note how the informatic model is used to assign a special status to DNA as that which initiates activity in the cell (Oyama, 2000; Lewontin, 1991). It identifies DNA as the uncaused cause of what goes on around it in the cell nucleus: the original message, the recipe that will be followed, the instruction that will be obeyed, the program that will determine, or, in the extreme formulation, the controlling or master molecule. What remains—now implicitly defined as the environment of the DNA—is then identified as something that is affected by DNA, changed at the instigation of DNA, but not itself a source of activity or change. The passive status of this environment is then taken for granted in the course of dramatic narratives that give DNA the lead role.

All this must be contrasted with the kinds of naturalistic account that are the norm in chemistry, where molecules are simply molecules and there is no justification for identifying some as possessed of fundamentally different powers from others or for anthropomorphizing the role of some and not others. The constituents of molecular systems demand understanding as fundamentally the same sorts of objects, and cells are such systems and never mere repositories of DNA. As well as DNA, every cell will also contain a range of RNA transcripts, proteins, and many other substances, all of which are involved in its functioning. Even water plays its part, and indeed its remarkable chemical properties give it a vital role in the functioning of the cell. At the molecular level, the cell is an elaborate chemical system with a self-stabilizing and boundary-maintaining mode of operation to which most of its constituents make essential contributions. Nobody would wish to pretend that DNA could function other than as one component among others in a chemical system of this kind, dependent on the parallel functioning of other parts of the molecular system. Yet there is no recognition of this in informatic narratives that use anthropomorphic metaphors to focus attention on the information DNA carries and, if taken literally, falsely identify the DNA molecule as the initiator and controller of the chemistry of the cellular system.

We ourselves are strongly committed to naturalism, and given this commitment it goes without saying that where informatic/anthropomorphic models and chemical/naturalistic ones clash, we would question the standing of the informatic account. Even here, however, we would not insist that its anthropomorphic metaphors are necessarily to be condemned. Many critics regard these metaphors as falsely implying a form of determinism involving DNA uncomfortably close to the objectionable genetic determinism they associate with classical genetics. But we should not criticize metaphors for being false or misleading if taken literally. Taken literally all metaphors are false, indeed taken literally they are no longer metaphors but mere mistakes. But accepted for what they are, as metaphors, they may structure and inspire the imagination to great effect. With metaphors and models it is the mode of use that is crucial, and the mode of use, if anything, that should be criticized. It is always possible that users both derive great advantages from deploying their models and metaphors and take care that misleading aspects of them are ignored or set aside.

This seems largely to have been what has happened in technical contexts when anthropomorphic metaphors have been used to describe DNA. The resulting accounts of DNA as code or program have proved invaluable to researchers, both as sources of creative inspiration and as means of narrowing and coordinating perception. And despite the charge that they convey a false determinist vision of what DNA does, and downgrade the importance of the context surrounding it, their employment in research seems to have done little harm. Evidently, researchers themselves have a fair idea of which aspects of their metaphors are valuable, which are liable to mislead, and which are palpable nonsense.[14] The message would appear to be that the use of anthropomorphic models and metaphors in genomics is perfectly acceptable if they are employed with proper care and probity. However, beyond their original setting in primary research it may be that their employment is more likely to create problems, whether because of the unworthy objectives of those who deploy them in these new contexts, or because they are liable to lead to genuine misunderstandings among audiences unfamiliar with their circumscribed uses in their original contexts.

For researchers, informatic models of DNA have been invaluable heuristically as ways of directing attention to specific, selected features of the underlying molecular models on which they are superimposed. They are of more practical use than the more complex molecular models that they serve to structure and simplify, but the more complex models have the greater ontological authority: they are reckoned to be "closer to reality." The relationship of informatic and molecular models is thus in some ways like that between the molecular models and DNA itself. Informatic models further simplify and structure the more complex molecular models that themselves give structure and relative simplicity to our perception and understanding of the material we identify as DNA.

14 Even the metaphor of DNA as program need not blind us to context: a program needs the right computer in which to function after all, so it is even arguable that a program metaphor should sensitize us to the importance of context.

And even though the molecular models have considerable ontological authority, DNA itself has still more. DNA is stuff, and we are able to relate directly to it and gain knowledge of it as such—where to find it, how to separate it off, how it behaves as material, and so forth. Knowledge of DNA has to be validated by reference to the stuff, and if work on the stuff creates problems for our current models of its structure, then in the last analysis it is the latter that need to be changed.

In a world in which our encounters with DNA are dominated by informatic metaphors, it is important to grasp imaginatively that DNA is indeed a tangible material substance. The organic bases that figure large among its constituents, thymine, cytosine, and the rest, are available as industrial chemicals, in fifty- or hundred-kilogram kegs. It contains sugars, similarly available, and phosphates, as found in fertilizers or in Coca Cola. DNA polymer could itself be manufactured in quantity if anyone had a mind to do so. In practice it is commonly studied and manipulated in minute amounts, but we do eat vast amounts of it and we modify it every time we cook a meal; it will be a long time before DNA-free food hits the supermarket. In the context of scientific research DNA has to be treated as a material substance even if this has been pushed into the background by the rise of an in silico research practice that investigates it in the form of immense four-letter sequences stored in computers. And while it is relatively uncommon to find DNA function being addressed from the perspectives of microengineering or the science of materials, a significant amount of work with this orientation does exist, and there has long been a body of existing knowledge for it to build upon. DNA has been recognized as a material constituent of the cell nucleus for well over a century.

The importance of the gross properties of macromolecules of DNA as they exist in cellular systems is easily illustrated by citing just two of them, which stand in intriguing contrast to each other. On the one hand, there is the extraordinary chemical stability of these molecules, which makes it possible for crucial information to be stored in a cell by just one of them, without any backup. On the other hand, there is their extraordinary spatial flexibility, which allows DNA filaments to take many forms, including packed down forms spooled around protein cores for storage and transportation, and unwound or partially unwound forms that facilitate replication and gene-expression. This balance of structural stability and spatial flexibility is crucially important in accounting for the extraordinary dance of the chromosomes that accompanies the division of the cell, whether in meiosis or mitosis. DNA functions in these systemic changes by virtue of its gross material properties, as well as by virtue of the information contained within it. Note as well that reference to context is also essential here, if we are to understand how the chromosomes are extended spatially in their surroundings, and to how other molecules in those surroundings are related spatially to them and thereby able to interact with them. Even the routine transcription of the information in coding DNA is now understood partly in terms of spatial relations: to take just two simple examples from many, base sequences may be either readily accessible and surrounded by relatively uncrowded space, or coiled away closely packed and next to impossible to transcribe; or again, some sequences may be close to "unzipping points" on the DNA filament, and transcribed more frequently because of that than those that are harder to reach.

A distinction has also long been made between two forms of chromosome material, heterochromatin and euchromatin, identified by the staining techniques employed to make chromosomes visible under the light microscope.[15] The darker stain of heterochromatic regions suggested the presence of a greater density of material and was later interpreted as a sign of tightly condensed DNA not readily available for transcription. And the very low levels of transcribing activity observed in them was one reason that there was no great concern when they proved very difficult to sequence during the great genome projects. But it is now recognized that heterochromatin has crucial roles as molecular material; these are increasingly being studied. Thus, at the center of each (eukaryotic) chromosome is the centromere, a dense heterochromatic section, which is the point at which homologous pairs of chromosomes attach to one another during mitosis and meiosis, and there are other heterochromatic regions at the commonest sites of the attachment and crossing-over that occurs during meiosis. The material at the very ends of chromosomes is also fascinating when considered in mechanistic terms. Here are the telomeres, repetitive sequences that serve the important function of compensating for the fact that replication cannot reach the end of a strand being copied, so that the tip of the chromosome is lost in each mitotic cell division. This tip is one of the telomere repeats, and it is currently conjectured that the number of telomere repeats functions to define the

15 Not all chromosomes, or things now called chromosomes, manifest the two forms of chromatin originally observed. Most bacterial chromosomes, for instance, lack the histones that form the protein core, and hence do not include heterochromatin, although what was once thought to be a sharp distinction between prokaryote and eukaryote chromosomes in this respect is becoming increasingly difficult to sustain (Bendich and Drlica, 2000).

life span of the organism. Finally, a single-stranded stretch of DNA called the T-loop connects the ends of the two strands of DNA in each chromosome, which prevents the end from being identified as a break by the DNA repair machinery.

Chromosomes and their DNA need to be understood as material things all the time, even when they are transferring information. DNA transcription is itself a physical process. Informatic metaphors may usefully be applied to it, but to understand, for example, how rapidly and accurately transcription proceeds it is necessary to remember that it is a process involving specific materials. And the same perspective is desirable when considering how transcription is regulated and switched on and off. A spectacular example of the permanent "switching off" of the transcription of DNA has long been known to geneticists, if not always under this description. Early in the embryonic development of human females one of the two X chromosomes in every cell changes in a way that deactivates almost all of its DNA. The particular chromosome inactivated in any given cell appears to be randomly selected; it may derive from either parent. Just a single one of the two original copies of the X chromosome remains active in any given cell, and indeed only one "normal" X chromosome can be observed in any cell in an XX human female. Instead of the second, a more compact object called a Barr body can be observed through the light microscope—so readily observed, in fact, that it was once used in attempts to confirm the sex of women athletes competing at the Olympic Games. The Barr body turns out to be the DNA of the entire second X chromosome, packed down into a small volume and coated with protein, which makes most of its DNA sequences impossible to access.[16] Since the inactivation is a heritable characteristic of the cell, a curious consequence of this phenomenon is that human females are genomically and genetically mosaic, with different X chromosomes operative in different clusters of somatic cells (Migeon, 2007). Even so-called "identical" (or monozygotic) female twins are mosaics and therefore not in a true sense genomically identical, due to different X chromosomes switching off in different somatic cells dur-

16 It is currently conjectured that all but a small proportion of the chromosomal DNA is taken out of commission in this way because there would otherwise be overproduction of the proteins the X chromosomes code for and a consequent imbalance in cell chemistry. As an indicator of sex, the Barr body has the disadvantage of occurring in the somatic cells of XXY and XXYY humans. As a basis for the administrative ordering and sorting of athletes it has, of course, far more serious disadvantages and thankfully its use has been abandoned.

ing ontogeny.[17] This does not, however, appear to be widely recognized by authors of twin studies such as that discussed in chapter 5.

The regulation of transcription does not always involve the kind of drastic change involved in the creation of a Barr body, or one so difficult to reverse. Many other things are known to affect transcription, and to regulate protein production with different degrees of sensitivity and reversibility. One is interference with the promoters of transcription through reversible DNA methylation, which we shall briefly discuss in the next chapter. Another is the presence of RNAi: double-stranded RNA complementary to a particular DNA sequence has been found to interfere with its transcription and to be capable of completely inhibiting it. For something with such a highly specific, sequence-linked effect an informatic form of understanding is obviously to be sought for, but at this writing the way that RNAi produces its effect is not known and the nature and extent of its role in natural systems are unclear. It represents an immensely important scientific advance, honored with a Nobel Prize, but at present it is an advance not so much in understanding as in techniques and powers. Researchers can now switch off precisely those "genes" in which they are interested using RNAi, and it is becoming a widely used and greatly valued tool both in scientific and in medical research.[18]

Genetic knowledges and their distribution

Let us recap once more. As researchers in classical genetics became more and more convinced that genes were real objects, the role of those objects in the production of the corresponding phenotypes cried out for investigation. With recognition of the importance of DNA, and the advent of the potent new techniques of molecular genetics, such investigation became a practical possibility, and the aim of closing a long-recognized

---

17 Mosaicism has manifest consequences at the level of the phenotype. In cats it has long been recognized as producing variations in coat color, and notably the all-female tortoiseshell coat. In human females it has significant implications for some aspects of female health (Renault et al., 2007; Migeon, 2007).

18 An oral contraceptive using RNAi is under development. The plan is to inhibit the production of a protein made only in developing ova, which is vital to their penetration by sperm. The proposed contraceptive will apparently have no effect on any other human cells, implying safety in use and a lack of side effects, and its effects will cease once its use is discontinued, since stored ova are unaffected.

gap in our knowledge became a realistic one. But as the new techniques transformed research, so they also changed the nature of the problems to be investigated, which increasingly came to be formulated as problems directly concerning the structure and function of DNA and the associated molecular systems. Chemical molecular models came into prominence, decorated with metaphorical redescriptions of DNA as information, or code, or program; and the reality of genes, so recently accepted, became open to question once more. Informatic accounts of genomic DNA allowed gene-like functions to be attributed to the substance directly, in ways awkward to translate into talk of molecular genes. Work on genomes as material entities further undermined not only the standing of genes as real objects but the ontological authority of classical genetics more generally. And currently research on the many other molecules in cellular systems is encouraging a further drift of ontological authority wherein even the attribution of gene-like functions to DNA molecules is being recognized as a radical simplification of what is really going on.

The advent of the new molecular approach, however, needs to be understood within a larger context still, that of the overall system of technical and intellectual division of labor in society. Research in molecular genetics and genomics is just one of the many forms of specialized technical activity that make up this system. And only in the narrow context of that one form of activity are molecular models elaborated in rich detail and their salience and value fully appreciated. A detailed understanding of these models cannot be sustained in the everyday world, and even other technical specialists, including researchers in other scientific fields, are often obliged to relate to them in simplified and stereotyped forms. Medical and legal professionals, for example, who have to take account of newly acquired knowledge and to convey its implications to patients or clients, will not necessarily understand that knowledge in the way that its producers and core users do. They may even find it necessary to revert to earlier models, structured around the schema of classical genetics, not merely to communicate with others but to assist their own understanding. And even in the realm of genetics and genomics research itself, for all that DNA is everywhere at the forefront of attention, the level of interest in how it functions varies considerably.

What is being put forward here is an impressionistic picture; there is as yet no detailed empirical study that serves to confirm the view on offer. We are, however, now beginning to acquire an understanding of how classical Mendelian genetic knowledge was distributed and put to use, and it offers some support by analogy for what this book proposes.

Mendelian genetics became established in the first half of the last century as the basis of research practice in a number of specific academic settings and in some applied contexts. It was, however, widely respected beyond those settings and increasingly acquired ontological authority beyond its context of direct use. Although there were those who were opposed to it, many specialists aspired to follow its methods, and where they could not and were obliged to persist with alternatives, as was often the case, they tried to make sense of what they were doing ex post facto, in a Mendelian frame. Thus, while it was extremely difficult for farm animal breeders, or clinicians and medical researchers, or psychiatrists, actually to do Mendelian genetics, it enjoyed high standing among them and, sometimes in modified versions, was used by them as an authoritative rationalizing resource. And by being deployed in this way, particularly by specialists who had dealings with the public, Mendelism also gained a toehold in the everyday culture, again often in modified forms, and again coexisting alongside different bodies of knowledge.[19]

Returning to genomics and the new genetics, it seems plausible to suggest that what we now have is a social distribution of "genetic knowledges," different sets of shared beliefs all of which are authoritative to some degree in some contexts. Different knowledges, in this sense, appear to be established in different settings, among different groups or subgroups of people, who put them to different uses alongside other related beliefs and practices. Among these are sophisticated molecular models strongly embedded in the practice of specific groups of laboratory researchers and largely confined to them; a variety of informal theories of ancestry and inheritance, some of which refer to genes and some of which make no mention of them, established in the context of everyday life; and in settings where links and mediations between specialists and their various audiences occur, older models of classical genetics and other ways of conceptualizing genes functionally continuing to have significant roles.

Within any division of technical and epistemic labor, however, there has to be a distribution of epistemic and cognitive authority existing in parallel to the distribution of knowledge. Few of us today invariably use whatever "knowledge" first comes to mind to guide our actions. We prefer to look to expertise much of the time and to give expert knowledge priority over our own initial beliefs and ideas in those domains

19 These claims were vividly illustrated in papers given by Hans-Jörg Rheinberger, Staffan Müller-Wille, and others at the conference on Cultural History of Heredity, Exeter University, December 2006.

where expert authority is recognized. The merits of a division of labor constituted in this way as a means of carrying and utilizing vast amounts of knowledge and competence are currently underestimated by many commentators. There is a tendency to emphasize instead how impressive the knowledge and powers of understanding of the public (or publics) are and to minimize the differences between ordinary citizens and elite specialists. But a better way of attacking elitism in this context would be to emphasize how relatively little experts know rather than how much ordinary citizens know. Certainly, outside their own special domains, which can be extraordinarily narrow, experts are in the same boat as the rest of us, and if ignorance and the need to defer to the knowledge of others are to count as indignities, they suffer them as much as anyone else. As J. Robert Oppenheimer (1958) remarked of mathematics half a century ago, the problem is not that nonmathematicians no longer know it but that mathematicians no longer do either, each of them being familiar only with a minute amount of the knowledge of the entire subject.

No doubt the current distribution of genetic knowledges is the product of diffusion processes, with the models now used by specialists slowly becoming more widely known, just as had occurred previously with classical genetics. But diffusion is shaped by the interests of recipients, who tend to acquire new knowledge according to their needs, and to modify it as they acquire it to conform with what they take themselves to know already and with their practical concerns. Thus, the models used by specialists in genetics and genomics, however useful, may remain unfamiliar elsewhere simply because most of us trust these same experts with their application and see no need to learn more of them. And where the expertise of these specialists does need to be communicated to others, to carriers of genetic diseases for example, the older, simpler schemes of classical genetics may often be preferred for that task. Accordingly, much of the current distribution of genetic knowledges may well long endure as a functional arrangement in which different schemes are adapted to different roles and uses at different social locations. In particular, despite their evident limitations, accounts that refer to genes, and even to "genes for," are likely to continue in use, and even to thrive and multiply, both in society generally and in technical contexts where they have instrumental value. And indeed, such is their instrumental value that the remaining chapters of this book will continue to have recourse to them.

Just as there is a division of labor within which genomics and molecular genetics are components, so too is there a division of labor within these fields themselves, and variation in the distribution of knowledge across them. The dominant approach in that internal division of labor long made use of informatic models focused overwhelmingly on DNA, even though chemical/molecular models carry the greater ontological authority and the importance of the cellular systems in which DNA functions is generally recognized. We have already suggested one reason for this: the prediction and/or explanation of difference was long the thematically dominant objective of genetics, and informatic models of DNA highlighted those differences between strands of DNA that are most salient in predicting differences of other kinds. But molecular genetics is now concerned with the controlled production of differences, as well as with predicting them, and we need to elaborate what we said earlier accordingly and recognize that DNA now serves not only as our major difference predictor but our major difference maker as well.

Consider the possibilities for intervening into and exerting control over developmental processes. It has long been recognized that the outcomes of these processes may be profoundly affected by intervention at their beginning, by careful selection of the sources of the initial gametes. This is the technical basis of the selective plant and animal breeding practices engaged in since time immemorial, practices that led not merely to "improved" individuals of given species but to the production of transgenic creatures as well, like mules. Today, of course, the production of transgenic creatures is a matter of routine genetic engineering, but current technology is nonetheless almost as limited as the old in one crucial respect. While many kinds of molecule are essentially implicated in ontogeny, only one of them even now can be acted upon systematically to engender sought-after differences, and that one only by intervention at the very beginning of the process. Small wonder then that DNA, and particularly the information implicit in the ordering of its four constituent bases, has attracted so much attention in specialized technical contexts: this is currently the best place to look if one's interest is in difference making.

It is worth pausing for a moment here to consider how this affects our understanding of the causal powers of DNA. In chapter 1 we noted how DNA is often identified as the cause of, for example, a phenotype having one trait rather than another, even though a whole nexus of causes is actually involved in the production of phenotype and trait. We nevertheless went on to defend those who cite DNA as "the cause," noting that some causal links may particularly interest us because they are difference makers, and that there are legitimate forms of causal discourse within which we may count only these as causes and treat the rest of the causal nexus as a background of necessary conditions. What we can now add is that the difference-making causes that particularly interest us are often those where we ourselves are involved in the difference making and can manipulate the difference maker to bring about effects congruent with our wants and desires. DNA is currently unique among molecules in cellular systems as far as our ability to manipulate it to bring about such effects is concerned. Accordingly, we now tend intuitively to identify DNA as highly potent causally, but water, for example, as a mere condition necessary to its operation, even though water is actually just as much implicated in the relevant causal nexus as is DNA.[20]

Let us return to the distribution of different genetic "knowledges." So far we have related it to the existence of a division of labor wherein specialists have different technical concerns and even the knowledge of genetics carried by or aimed at general audiences is related to technical and instrumental interests of some sort. But important though they are, narrowly technical interests and objectives are not the only ones involved here. Knowledge may be taken up and deployed for any number of reasons and put to every conceivable kind of use. Its acquisition may at times be moved by a general intellectual curiosity, and with genetics/genomics as with other sciences this curiosity will tend to focus on changing theoretical frameworks and models. These are cultural resources that can be deployed in the context of everyday discourse and woven into coherent narratives, whether in attempts to make overall sense of the world or in efforts at self-understanding. And of course these same resources may also be deployed in the creation of persuasive rhetorics, with specific social, political, and moral objectives that will vary according to where those who deploy them are located in the social order and what their specific concerns and interests are.

Let us clarify why we speak of theories and models as cultural resources. In commentaries on the rise of genomics and the new genetics a key question tends to be whether or not current stereotypes of genetic determinism are reinforced or undermined by these new sciences.[21] One view here is that genomics offers a vision of basic biological processes

20 This emphasis on the ability to make a difference resonates with an important theme in philosophical discussions of causality, recently reinvigorated by the influential work of James Woodward (2003).

21 The actual question tends to be whether genomics supports or undermines the deterministic explanations of "classical genetics." It incorrectly assumes, that is to say, that explanation in classical genetics was pretty much identical to the simplistic genetic determinism often encountered today in our own media. We discuss this media genetics in chapter 5 as "astrological genetics."

properly attentive to context and environment and seeks to provide a holistic understanding of the structure and function of genomes; to that extent genomics implies a rejection of determinism and of reductionism as well. But another view is that a determinism analogous to genetic determinism is evident in genomics itself and especially in its informatic models and metaphors. Proponents of this second view take accounts of DNA sequences as programs or sets of instructions to imply that they determine cellular processes, and they criticize the accounts accordingly as stand-ins for rather than alternatives to hierarchical explanations in terms of determining genes. Their own preference is for a more democratic vision of the polity of the cell, in which DNA as determinant is reconceptualized as just one of the many functioning components of a larger interacting molecular system.

Our immediate purpose in juxtaposing these two apparently conflicting accounts of the implications of genomics is not to criticize either of them specifically. What we have here are not a correct and an incorrect account, but two different interpretations of genomics and its implications. And there is, in general, no indefeasible way of deciding between such interpretations, whether of genomics or indeed of any science. It is misleading to contrast "the" implications of genomics with "the" implications either of current varieties of genetic determinism or indeed of traditional classical genetics. Like all metaphors, models, and theories, those at issue here have no unequivocal implications, certainly no unequivocal social implications, and are imbued with significance only through human artifice; that is, through being interpreted and modified in use so as to take on a particular significance. This is why we ourselves prefer to speak not of the implications of genomic models and metaphors, but of their availability as cultural resources.

As far as the general social significance of these models and metaphors is concerned, therefore, what they provide is opportunity. They are pristine resources in a setting where resources previously drawn from the biological sciences have been very heavily modified over time for use in the wider context. For over a century, the two authoritative biological theories traditionally cited in works addressed to general audiences were those of Darwin and Mendel. What these theories "really implied," or even what these theories really asserted, it is hard to say. Historians' readings of the key texts vary remarkably. And the theories, if such there were, have since been repeatedly redescribed and reinterpreted, loaded with anthropomorphic and teleological excrescences, and run together in narratives from which (questionable) moral and evaluative implications appear to flow even more readily than empirical ones.

The molecular models of genomics and the new genetics, in contrast, have not yet been extensively coupled to the moral or evaluative concerns of wider audiences, but in the biological sciences they have rapidly acquired ontological authority as their technical merits have become evident. Through recourse to these models it is currently possible to construct widely intelligible accounts of ourselves and our relationships to other living things that are genuinely (and admirably) naturalistic and materialistic, even as it has become more difficult to construct such accounts from the schemas that preceded them.

Of course, we cannot assume that this will remain the case in the future. The supply of designer variants on the chemical-molecular models of genomics could rapidly increase, and they could increasingly be adapted for the kinds of nontechnical use to which modified Darwinian and Mendelian theories have so often been put, a particularly depressing thought given how frequently these nontechnical modes of use have been straightforwardly disreputable ones. It would be disingenuous of us to claim, however, that only elaborated and modified versions of scientific models and theories have social and political utility. Even if the molecular models of genomics are "fundamentally," "in their basic form" nonanthropomorphic and nonteleological, this by no means makes them intrinsically unsuitable for anything but narrowly technical forms of use. By ancestry, they stand among the cultural resources of materialism, and their recent role in the molecularization of genetics and the emergence of genomics could be made out as a success for a naturalistic and materialistic conception of nature that has long itself been regarded as a contestable one. The materialist tradition has indeed itself long existed as a treasure chest of cultural resources, constantly being interpreted in new ways, presented in new guises, attacked on new grounds, and imbued with different moral and political "implications" by figures as different as Karl Marx and Francis Crick.

3

Genomes

What are genomes?

According to a recent authoritative textbook, "Most genomes, including the human genome...are made of DNA...
but a few viruses have RNA genomes" (Brown, 2002,
p. 4). The textbook goes on to say that almost every somatic cell of the human body has two copies of the genome, contained in twenty-three chromosomes, while
human gametes or germ cells have just one copy.[1] This
is an example of what is currently the commonest way
of defining what genomes are, and we have no quarrel
with it. It takes (human) genomes to be "made of DNA,"
identifies the DNA in the twenty-three different chromosomes in a cell nucleus as being the DNA in question, and
notes the presence of two copies of the genome in every
somatic cell. This will be how we ourselves shall understand genomes, although we prefer to speak of each cell
having one or two genomes within it, in the form of the
DNA in its chromosomes, rather than of its having one

---

[1] Brown is here referring to the *nuclear* genome. He later
mentions the *mitochondrial genome* and the numerous "copies"
of that that exist in organelles in every cell.

or two "copies" of "the genome."[2] This serves to give greater emphasis to what Brown's definition already makes clear, that genomes are "made of DNA," and explicitly acknowledges what follows from this: genomes should be understood as material objects—the material objects pieces of which are cloned and sequenced in genomics laboratories.

Brown's definition builds upon our existing empirical knowledge and in particular on our knowledge of the chromosomes that under suitable conditions may be observed in cell nuclei.[3] Indeed, there is a close connection between current ways of representing genomes and earlier ways of representing a complete set of chromosomes as a karyotype or karyogram. The connection between how we understand genomes and how we understand chromosomes is particularly prominent in Matt Ridley's (1999) best-selling book, *Genome: The Autobiography of a Species in 23 Chapters,* where each of the chapters focuses on one of the twenty-three (pairs of) chromosomes. At the same time, however, Ridley's discussion highlights the other major historical link that has conditioned how genomes are currently thought of. Each chapter selects a gene from the relevant chromosome for discussion, implying that the genome is largely to be understood as an ensemble of genes strung out along the chromosomes. The main theoretical notion from which "genome" descends is of course "gene," and it is not uncommon to see the genome of a species or organism defined as the complete set of its genes. The Human Fertilisation and Embryology Authority's website, for instance, defines "the genome" as "the basic set of genes in the chromosomes in any cell, organism or species."[4] Unfortunately, this definition inherits all the difficulties with the concept of gene outlined in the previous chapter. And in the case of "the human genome," it is a definition that

2 Genomes seem to be thought of as material objects rather like books are, and just as several libraries may have a copy of *The Wind in the Willows* with each copy of the book being a book and the book with that title, but with nothing external to the copies that the copies are copies of, so it may be with genomes. Philosophers discuss the confusion liable to arise here as caused by a failure to recognize the "type-token" distinction. In genetics and genomics this confusion is ubiquitous, and references to "copies" are markers of where it is most likely to occur. Fortunately, it causes few difficulties in practice.

3 Or just cells, when these lack nuclei. We shall discuss the special features of prokaryotic genomes below.

4 See http://www.hfea.gov.uk/en/312.html. Even in Brown the same notion is present, with his glossary at variance with his main text in defining "genome" as "The entire genetic complement of a living organism" (2002, p. 525).

encompasses only a small amount of the nuclear DNA, which might be thought unfortunate. At any rate, we prefer to stay with a definition that includes all the DNA of the chromosomes and that avoids making the concept of a genome hostage to the fortunes of the gene concept.

A third form of definition, and the one that constitutes the main challenge to the policy of defining genomes as material objects, conceptualizes them in abstract and fundamentally informatic terms: the genome of an organism is all the information carried by the four-letter sequences of its DNA. Searching the Internet for definitions of "genome" provides a good mix of informatic and material definitions, with a few that attempt to have it both ways. This stands as testimony to the dominance of the informatic perspective in genomics and to the extent that informatic accounts have been disseminated into public consciousness. When the completion of the Human Genome Project (HGP) was announced, what had been achieved was a sequence of letters, A's, C's, G's, and T's, that was widely held to contain information, sometimes all the information, about the human organism. Familiar metaphors of the human genome, as blueprint, map, program, and so on, refer in somewhat different ways to this putative sequence and the information it contains, downplaying perhaps how the HGP was a project to sequence something "made of DNA."

We have already discussed the informatic perspective in chapter 2 and noted how productive informatic metaphors have been in inspiring and coordinating research in genomics and molecular genetics. We ought to acknowledge as well that an informatic definition of "genome" can avoid some of the problems associated with a material one, even if it raises problems of its own. And finally we should admit that information theory at a very general level provides a perfectly reputable basis for the study of DNA.[5] The general scientific understanding of the transfer of information derives from the work of Claude Shannon and Warren Weaver (Shannon and Weaver, 1948), who speak of an information source, a channel down which the information flows, and a receiver at which it arrives. Very crudely, information flow allows a state of the source to reduce uncertainty in a state of the receiver. So for example if you don't know what day of the week it is, and I inform you that I know it is Thursday, your seven possible beliefs are reduced to

5 Although the picture of DNA as informational, and even the wider view of cellular process as calling quite generally for analysis in terms of information flow, has become quite widely entrenched, philosophers have for many years been skeptical of this concept (Oyama, 2000; Moss, 2003, ch. 3).

one. Analogously, we may know little of the structure of a protein and be undecided between many possibilities, but we will know a great deal more about it and discount most of the possibilities previously kept in play when we are informed of the DNA sequence from which it derived. And it is also possible to think of information flowing from DNA itself reducing the possible states of a protein-producing system, so that in an appropriate context it produces a particular protein.

Finally, as a complete contrast to the informatic definition, we mention a possible way of defining "genome" with very considerable advantages, which is not currently widely accepted. The dominant view of genomes is that they are objects made of DNA. But the actual material objects we encounter in the cell nucleus are made of chromatin, not DNA. In chromatin, DNA exists in association with various other substances including small RNA molecules and proteins, and in particularly close association with the histone proteins that provide something like a spool around which the DNA strands are coiled, and which thus facilitate the packing of DNA into the restricted space available in the cell nucleus. Small RNAs are known to be implicated in gene interaction and gene suppression. And histone proteins are turning out to be much more than inert packing materials: modifications to histones are now recognized as having important effects on the regulation of gene expression. Information can be obtained from these molecules just as it can from DNA. And if genomes are taken to be material objects containing information salient to the characteristics of the phenotype, there are evident advantages in defining them so as to encompass all of the material of the chromosomes. Perhaps, as researchers continue to use the term "genome," just such an extension of its meaning will imperceptibly occur. Then, genomes will indeed be discernible in every cell nucleus, as the materials that constitute its chromosomes, and the space defined by these materials will come to be understood as a "space in which genes happen" (Hughes, 2005).[6]

Why then do we ourselves prefer the more restricted definition given by Brown? It is actually not a matter of preference. Nothing empirical follows from how terms are defined and users may define "genome" as they wish, but users have already revealed their preferences here, and a book of this sort needs to respect them and cannot strike out on its own. We follow Brown mainly because his definition of genomes as material

6 This phrase might seem to raise again problems about genes. However, it is clear that Hughes's aim is to stress the word *happen*, and hence interpret genes as events in the genome, thus de-reifying them.

objects made of DNA is very widely accepted. Of course, the informatic definition is also widely accepted, and we preferred not to go with that alternative. One relevant consideration here was that at the level of everyday understanding, where accounts of genomes could eventually become very important, a number of suspect elaborations of the informatic perspective are in circulation that it is better to remain distant from.[7] But more salient still were the positive advantages of adopting a materialist/molecular definition in the light of likely future developments of the science.

As research in this broad area continues to become more and more molecular, and it becomes ever harder to convey what it involves in terms of genes, references to genomes may come to serve instead as the basis of accounts through which wider audiences stay in touch with what is going on. But this research is moving with extraordinary rapidity and in directions that few have been able to predict long in advance. Whether it will remain as strongly focused on DNA and its sequences as it has been is very doubtful. Whether indeed there will be a scientific field identifiable as genomics in a decade's time is an open question. We cannot be sure that any picture or theoretical representation of what is being studied is going to remain helpful for a significant period in grasping what is going on in these sciences and in the realms of experience they are exploring. But those representations that are most selective and that narrow perception the most strongly, and those that refer only to objects or data of kinds that have already been made visible, are perhaps at greatest risk of quickly becoming redundant. Representations of genomes as four-letter sequences are very much of this kind, whereas representations of genomes as chemical molecules are far less so, and an understanding of genomes as the material objects that the molecular models model is, if we may put it so, more open to the future still. For this reason an understanding of genomes as material things may well provide the best basis on which to establish relatively enduring new patterns of everyday understanding through which the significance not just of genomic but of post-genomic research may be followed and its broad significance understood from the outside. Whether research persists in

7 As we mentioned in chapter 2, there is a misconceived tendency to regard the genome as a repository of information sufficient in itself to determine, regardless of context or environment, the phenotype of the organism. In texts directed to general readers this has been associated with genetic determinism and the flawed and frequently pernicious arguments associated with that older position.

a broadly reductionist movement through the "omes" (transcriptomes, proteomes, etc.), or expands into the study of molecular systems or the biochemistry of cells, or looks increasingly to context and environment, or concentrates on ontogenetic processes, or even remains predominantly informatic, a grasp of genomes as material things could well continue to prove useful in making sense of it.

In any event, we have decided to settle for the material definition, and we can now look in a little more detail at the characteristics of genomes as material things. The first thing to say here is that genomes vary enormously in size and in structure just as the organisms in which they reside do, and that the sequencing of DNA derived from different species has strikingly displayed that variety. Indeed, genomics has been described as a new version of natural history, recording the features of specimen genomes of different species much as the features of specimen organisms were previously recorded. Less kindly, it has been compared to stamp collecting. And indeed as far as size is concerned it is hard to know what to make of the variation that has been encountered. For example pufferfish, *Takifugu rubripes*, have genomes one tenth the size of those of humans, whereas the single-celled protist, perhaps aptly named *Chaos chaos*, is reported to have genomes some four hundred times as large.[8] The failure of genome size to bear any intelligible relation to the complexity of the carrier organism remains a puzzle, sometimes known as the C-value enigma, and neither is the association of estimates of gene number with complexity clear-cut and well understood.

There is more to be said about how genomes vary structurally. The great divide in living things between prokaryotes, in which cells lack internal partitions, and eukaryotes, in which membranes separate off the cell nucleus and other subunits, is mirrored in two different forms of genomes. Eukaryotic genomes, including those of "higher" organisms such as animals and plants, are generally embedded in sets of linear chromosomes, whereas prokaryotes, of which bacteria are the most familiar instances, typically have most of their genetic material arranged in a single circular chromosome.[9] Most bacterial DNA, up to 90% in

8 Although this estimate is open to doubt. For more information on genome sizes see http://www.genomesize.com.

9 We say "most" because bacterial genomes are in fact highly diverse, and most bacteria contain small additional nucleic acid structures, usually also circular, called plasmids. It has now emerged, however, that some bacteria have several circular chromosomes, and some have one or many linear chromosomes.

some cases, is both transcribed and translated, so that representations of bacterial genomes do indeed make them appear very like beads of adjacent genes on a string. And since these were the genomes that were first studied and sequenced, it came as something of a surprise when "coding DNA" turned out to constitute only a small part of the material of the genomes of humans and other eukaryotes and a whole range of differences in how they were structured emerged. Indeed, several widely accepted generalizations mainly derived from work on prokaryotes, and the workhorse model bacterium *E. coli* in particular, proved not to apply to eukaryotes (Darnell et al., 1990). Prokaryotes are overwhelmingly the most common organisms, and for 80% of the history of life on the planet were the only organisms, so it would be entirely appropriate for a discussion of genomes to take prokaryotic genomes as paradigms. But consistent with the parochial, anthropocentric perspective to which we have already confessed, we shall give more attention for the moment to the genomes of multicellular organisms such as ourselves.

With regard to the "coding DNA" previously discussed as the presumed material substrate of genes, there is little we need add to what we have said already. As we noted, the proportion of the genomic DNA it makes up is much smaller in eukaryotes than in prokaryotes, even allowing for the promoters, enhancers, silencers and so on mentioned in chapter 2. And the internal complexity present in this material also varies considerably, with exons and introns known to be the norm in humans and many other eukaryotes but infrequent in bacteria, and with variations in the incidence and form of alternative splicing systems and several other elaborate organizational features of eukaryotic genomes an important topic for research.

It has long been a commonplace in discussions of eukaryotic genomes to remark on how small a proportion of their DNA—little more than 1% in the case of human genomes—actually "codes for" proteins, and to refer to the remaining "noncoding" sequences as "junk" DNA. As with many ideas in the area, this concept owes much of its popularity to Richard Dawkins (1976) and the idea that the genome was a space in which genes were competing with one another in an evolutionary process quite independent of the interests of the organism in which they were ultimately lodged. However, the whole concept of junk DNA is becoming increasingly dubious in the light of a growing body of current research. It is clear now that far more DNA is transcribed than used to be thought, and estimates for this proportion now run anywhere up to 90%. It remains a matter of conjecture how much of this transcribed but untranslated DNA serves a function for the organism, but it is increasingly believed that much of it does. One especially telling finding has been the presence of so-called ultra-conserved regions in DNA sequences previously thought to be junk (Bejerano et al., 2004). These are sequences that show almost no differences between distantly related species, for example humans and mice, which is taken as very strong evidence that they must serve vital functions. What these functions are remains unknown, but it is increasingly widely believed that a large number of the many diverse kinds of RNA molecule have vital functions in the regulation of cellular processes, although the detailed working of this regulatory layer remains very poorly understood.

We shall say a little more about the features of "noncoding" DNA sequences of no known function in the next chapter, in connection with the evolution of genomes, but for now we want to emphasize the word "known" in the phrase "no known function." For the question of how much of this DNA has as yet unknown functions is very much an open one. Of course, whatever else, junk DNA functions as part of the physical material of the chromosomes (chapter 2); and this gives it a role in defining the "space in which genes happen" and in locating other pieces of genomic DNA within that space and separating them from each other therein. But there remains a substantial proportion of the genomic DNA, replete with structure and pattern, the *informatic* functions of which remain obscure, if indeed there are any. Genomes contain many so-called *pseudogenes*, DNA sequences similar to those of genes but not transcribed, because they lack promoters or because mutations have made them unreadable. The existence of pseudogenes has prompted considerable speculation. It has been suggested that they are defunct fossils of genes active in some ancestor and also that they may occasionally be reactivated and hence have a continuing significance as latent genomic resources. More puzzling still are repeat sequences, which occur in human genomes in profusion, and in so many kinds and forms that even their basic taxonomy would take some pages to describe. Among these repeats are so-called transposable elements or *transposons*, pieces of DNA capable of moving around and of copying themselves back into the main sequence at another location, whether at a distance or adjacent to themselves. Since this back-copying is a possible explanation of the proliferation of repeat sequences, it has encouraged the view that many repeats could be genuinely functionless bits of genomic DNA along for the evolutionary ride. Were this the case, then here indeed would be truly selfish "genes" providing no benefits to their host organisms, but recent discoveries that some of the highly conserved sequences just mentioned are composed of transposons make this supposition increasingly implausible (Pennisi, 2007).

Having mentioned transposons, we conclude with one final argument on behalf of defining genomes as material objects. Regardless of the definition, variations among genomes must be recognized. Many of the variations we have discussed have been between-species variations, and even those that have not been could mostly be reconciled with the notion that each species has "its own" genome and hence with the custom of referring to "the human genome," "the mouse genome" and so on. We talk of "the mouse" as a species, after all, even though all mice are different, so why not "the mouse genome" even though all mice have different DNA sequences? In the last analysis, of course, it is a matter of convenience.[10] Much variability, then, can be handled within such a terminological convention, but beyond a certain point it is likely to become too awkward to deploy and too liable to mislead. As research on human and other genomes progresses, it is moving more and more away from the task of producing one basic reference sequence per species to the study of variation; and how far just one reference sequence continues in use, underpinning and informing the work, will be a matter of how the research turns out and how the researchers decide to conceptualize it. It is perfectly possible that there will prove to be so much more within-species variation than expected, and so many more instances of gross forms of variation in the DNA, in the genomes as we want to say, of particular organisms, that the reifying discourse which speaks of "the genome" of this or that species will fall out of favor.

Certainly, it is taken for granted even by those happy to talk of "the human genome" that the genomes of every individual human being are different. Genomes are resources for individuating humans as well as celebrating their unity. Current research is relating variations in individual phenotypical characteristics to genomic variation, largely at the moment to the SNPs that are the focus of the International HapMap Project and similar large-scale enterprises,[11] but increasingly to larger units of

10 It may also be an attractive basis for politically oriented discourse, as when "the human genome" is proclaimed the common heritage of humanity as a whole (Bostanci, 2007).

11 Single nucleotide polymorphisms (SNPs, pronounced "snips") are single-letter variations in the genome; they are being extensively catalogued, especially in the human genome. The International HapMap Project (for "haplotype map") is an exploration of the geographic distribution of such variants (see http://www .hapmap.org).

material.[12] Variations in other less localized genomic features, such as the numbers of repeats in repetitive sequences, have also been studied, and these are now used as markers of individual human identity, for example in databases being compiled by police forces and the military. But there is an enormous amount more to be learned about the extent and the importance of genomic variation among individuals, and a radically revised understanding of its significance is to be expected even in the short term, now that so much attention is being devoted to it. Not even this kind of work, however, takes the study of genomic variation to its limit. Just as it may be taken for granted that the genomes of every individual organism are different, so may it be taken for granted that the genomes of every individual cell of an organism are different. Here, however, is a topic that needs a section of its own.

The strange case of the epigenome

The DNA the Human Genome Project sought to sequence was that to be found in a single complete set of human chromosomes. It was thus roughly half of the DNA present in the nucleus of every human cell. And since each chromosome contained just one DNA filament, it could be described as a single long DNA molecule broken into over twenty pieces. This does indeed seem to be how genomes are most widely regarded by genomic researchers: a genome is the entire DNA filament embedded in the set of chromosomes found in the nucleus of every cell of an organism. The organism itself, if it contains billions of cells as humans do, will contain billions of genomes. Genomes should accordingly be regarded as material objects, embedded in the chromosomes that inhabit the nuclei of cells. Or so we have urged in the first part of this chapter.

Proper formulation of the materialistic perspective is not without its problems, of course. It is clearly incorrect, for example, to think of genomes as pieces of DNA floating about in cell nuclei, or even as molecules of DNA tangled up with proteins in chromosomes. Chromatin, the material of chromosomes, is not a mere mix of DNA and protein: these constituents are chemically combined in chromosomes and do not exist as separately identifiable components or, therefore, as discrete material objects. We could say that genomes are idealized objects and not actual pieces of material. Moreover, even the DNA separated out from chromatin dif-

12 One important group of variants increasingly being studied are the "indels," the insertions or deletions of lengths of sequence that caused some small problems to the Human Genome Project.

fers in important respects from genomic DNA as generally understood. The DNA that is "read" in the great sequencing projects is manufactured from natural DNA in ways that transform it chemically, although the transformed DNA is nonetheless believed to retain, and when sequenced to reveal, crucial structural features present in the natural material.

If the process of manufacture here merely involved the separation of DNA from protein with which it was very loosely linked and which had little effect on its functioning, then it might hardly be worth calling attention to. But the chemical links that have to be severed here may be strong and the conjoined protein material of very considerable functional significance. Moreover, in the course of preparing pieces of DNA for sequencing the DNA itself is significantly transformed.[13] Indeed, it is transformed to the extent that the four-letter text read off from it is in an important sense an artifact of the processes involved and not a record of something intrinsic to the "natural" DNA itself. Certainly this is the case if the four-letter text is understood as referring to a sequence of nucleotides corresponding to the four organic bases, adenine, cytosine, guanine, and thymine. Sequences of precisely this sort are not found in the DNA of human chromosomes, although they do provide us with valuable information about that DNA. It could be said (although it appears not to have been) that the four letters accurately identify the skeletal structures of the four bases, so that C, for example, identifies not cytosine, but rather the skeletal structure common to cytosine and a number of other single-ring pyrimidines. And indeed the sequence of four kinds of skeletal structure is knowledge of great value and importance, of something found both in natural DNA and in the DNA manufactured from it for the sequencing machines to chew through. Put another way, "cytosine," in the context of the standard representation of genomes, does not refer to a specific chemical species, but rather to a chemical genus, one of any of a number of more or less similar chemical kinds.

As far as fully specific organic bases are concerned, a considerable number (currently over twenty and rising) have been found in natural DNA. But nearly all of them may be regarded as variants on four basic kinds, consisting in different subunits attached to one of the four skeletal structures. Hydrogen, –H, is by far the most frequently encountered subunit of course, so much so that it is generally left implicit in representations of nucleotide structure, but other subunits (notably $-CH_3$, the methyl group, and $-NH_2$, the amine group) are also relatively common.

13 A growing number of techniques for DNA sequencing are currently being developed, but the general problems described here apply to all of them.

Again, this might hardly be worth mentioning if all the bases of a given kind, with a given skeletal structure, functioned in very much the same way, but they do not. They function very differently, and the differences are proving to be of great biological significance. The cases where this first became clear and evident involved the methyl group, $-CH_3$, and in particular the presence in natural DNA not of cytosine but of 5-methyl-cytosine at locations in the four-letter sequence marked by C.[14] Subsequently, however, the important role of other subunits has been recognized, and their presence and functional importance has been noted as variable components not just of natural DNA but of the proteins associated with that DNA in chromosomes. Currently, the extent of its methylation is recognized as functionally the most important of the variations of this sort in natural DNA and along with acetylation (involving $-COCH_3$) as important also in the histone proteins associated with DNA in eukaryotic genomes. The study of these variations and their effects is now sometimes called "epigenomics," and all the natural DNA of a set of chromosomes, complete with the variations just described, is sometimes referred to as "the epigenome." For the sake of simplicity, however, we shall discuss epigenomics in what follows as if the only variation involves the methyl group and cytosine.[15]

In natural human DNA, 5-methyl-cytosine, and not cytosine, is present in a significant proportion of the cases where the sequence text reads "C." Indeed, if it were a mere matter of frequency the base would merit its own letter, "M" perhaps, and DNA would be better imagined as a five-letter text. It would also be more consistent to give 5-methyl-cytosine its own letter. After all, 5-methyl-uracil has its own letter, T, even though this base, also known as thymine, has the same skeletal structure as uracil, U, so why should 5-methyl-cytosine not be similarly treated? The answer, of course, is that nomenclature is partly a matter of historical accident and partly a reflection of what interests scientists most and what they wish to highlight at a given time. Four basic letters allowed the development of the bioinformatic approach to DNA, and a fifth letter, U,

14 In practically all cases in mammalian genomes the 5-methyl-cytosine is actually located at a CG site. But in bacterial genomes this is apparently not the case, and hence in the light of what is to come in chapter 4 we perhaps should say that CG methylation is unusual.

15 "Epigenetic" and "epigenomic" phenomena have been understood and defined in very different ways at different times and the semantics of the terms remain unstable and rapidly changing. Even specialists, however, sometimes still think of "the epigenome" simply as a methylated genome.

highlighted an important difference between DNA and RNA: where thymine appears in a DNA sequence, uracil replaces it in the RNA sequence derived from it. As research interests shifted and broadened, both nomenclature and underlying schemes of thought were developed by the extension and elaboration of what was first established. Thus, 5-methyl-cytosine is currently treated as methylated cytosine, and there is currently much interest in methylation as if it were something that happens to or is done to genomes. It is easy to get the idea that genomes are natural objects that become methylated in certain circumstances, when in reality the nearest things to such idealized genomes we have experience of are the "demethylated" DNA filaments we specially prepare for sequencing. If there are material objects in the natural world the existence of which justifies our references to genomes, then those objects are "methylated" genomes, or epigenomes as they now tend to be called. It is indeed a very strange quirk of terminology that identifies DNA in or close to its normal natural states as part of an epigenome and DNA in a radically modified state as part of a genome, but that seems to be how things stand at the moment.[16]

There is currently great interest in the extent to which human DNA is methylated and in the pattern and distribution of the methylation. The genome has been mapped and sequenced, it is said, and the time has now come to map and sequence the epigenome. It is a call to action reflecting far more than the interests of researchers in laboratories bristling with expensive technology in search of something new to do with it. It is prompted mainly by what the consequences of methylation, and hence its biological functions, are now believed to be. C-methylation is now widely reckoned to reduce or to switch off DNA transcription, and hence control and regulate gene expression.[17] To map its distribution and reveal its mode of operation are accordingly vital steps toward an understanding of how DNA actually functions in the molecular systems of the cell, and in the chemical economy of the organism as a whole.

16 References to "the epigenome" now common in the literature amount to an even more problematic reification than do references to "the genome,", since all the kinds of somatic cells have different epigenomes, as we shall see. On the other hand, maps of epigenomes make an interesting contrast with representations of genomes in that they are usually maps of the probability densities of methyl groups.

17 We shall not discuss how methylation inhibits gene expression. The current view seems to be that the promoter regions and not the "coding DNA" itself are affected. But insofar as we can tell it has not yet even been decisively established that DNA alone is involved in the relevant changes.

Knowledge of cytosine methylation and its significance must be recognized as tentative, conjectural, and liable to rapid and radical change as work on these topics continues. Even so, it is clear that this is a truly fascinating research field and that even its provisional findings are raising challenging questions and undermining long-accepted assumptions. Thus, it has largely been taken for granted since the advent of classical genetics that while the traits of the phenotype are a consequence of the specific nature of the genetic material it inherited, they are not affected by the origins of that material. But experimental findings have now convinced researchers that this is false. In normal sexual reproduction one chromosome of each pair derives from the male parent and one from the female, but it has been possible in mice to produce zygotes in which both chromosomes derive from one parent. Even though both these chromosomes are perfectly normal genomically, that is in terms of the four-letter sequences of their DNA, and all that is manifestly unusual is the path by which they have reached the mouse zygote, radical consequences are apparent. In many cases, implantation and/or development fail to occur. In others, phenotypical variations (generally pathological) are observed that cannot be accounted for along traditional lines. The evidence currently suggests that responsibility lies with methylation patterns, which create differences in the sets of DNA sequences initially available for transcription on chromosomes that originate from different parents. Both of these sets are essential for normal development. These methylation patterns, it is believed, are *imprinted* on the chromosomal DNA during gamete formation in the parents. And since they are specific to the sex of the immediate parent, it is believed that the process of imprinting occurs anew in every generation. During gamete formation, in other words, certain existing methyl groups are stripped from the DNA, and it is then remethylated with a pattern specific to the sex of the parent. This process is sometimes referred to as "reprogramming."

Genomic imprinting remains an inadequately understood phenomenon and it may be that much more is involved in it than DNA methylation. More is bound to be learned about it as research on a range of model organisms proceeds, but we can already be reasonably sure that what is learned will be relevant to humans.[18] Several rare human genetic

18 Thus a recent collection reviewing research on DNA methylation, published as a special edition of *Nature Reviews* (October 2005; www.nature.com/reviews/focus/dnamethylation), lays immense emphasis on its association with cancer and the likely utility of the research in the detection and treatment of cancer in humans.

defects are known to manifest themselves differently according to the parental chromosome in which they occur, and there are strong grounds for believing that the explanation lies in the different methylation patterns on the two chromosomes. The most widely known case is probably that of the deletion that when inherited from the father gives rise to Prader-Willi syndrome and when passed on by the mother to Angelman syndrome. There are also very rare pathologies where the involvement of imprinted methylation patterns is strongly implied but no genetic/genomic defect of the traditional sort exists at all. Occasionally, individuals have a complete and normal set of chromosomes but have derived both of a particular pair from the same parent. The condition is known as uniparental disomy and has been found to be accompanied by serious phenotypical abnormalities. Again, imprinting is the obvious explanatory hypothesis.

Parental imprinting on the two genomes of the human zygote may set the initial state from which the growth and differentiation of the zygote begins, but such imprinting can be no more than a starting point. Methylation patterns are bound to vary as ontogeny proceeds. Even if we view a complete human being merely as a collection of cells and ignore its high level of cellular organization, we still have to face the fact that well over a hundred kinds of terminally differentiated cells make up the human phenotype and that the key differences between them lie in the proteins they produce. This in turn implies differences in the DNA that is available for transcription within them and hence, according to current thinking, in their methylation patterns. Thus, a very important feature of the process of ontogeny is the transformation of the DNA of cells from one pattern of methylation to another, with many such epigenomic patterns eventually manifesting themselves as different cell types. Clearly, methylation is not permanent, and both methylation and demethylation must be understood as essential processes in the context of the life cycle. At the same time, however, methylation patterns do typically have some considerable stability and are not easily transformed. Once the first of a new cell type appears in the course of ontogeny, its characteristic mode of protein production, and hence by inference its characteristic methylation pattern, persists in all the cells deriving from it. Methylation patterns are highly heritable in cell populations reproducing by mitosis.[19]

19 This is the conventional way of noting that the characteristic may pass from generation to generation without change and it is clear intuitively what "highly heritable" means here. We discuss below the question of whether methylation patterns are heritable from parent to offspring. "Heritability" is, how-

Precisely because they have called so much longstanding biological orthodoxy into question, epigenomics and epigenetics have given rise to controversy and ideological conflict just as the fields from which they derive have done. The apparently esoteric technical question of how stable methylation patterns are has actually been perceived as of great moral and political interest ever since the existence of epigenetic inheritance in somatic cell lines was first reported. This was because it was a case of the inheritance of an acquired characteristic, or was regarded as such a case, and as a matter of history an important part of the fight to establish the dominant Darwinian/Mendelian perspective in modern biology had been against proponents of this Lamarckian kind of inheritance. Even in the case of somatic cells and mitotic reproduction, reference to the inheritance of acquired characteristics seems to have evoked collective anxieties, although of course little actually hung on this; the key doctrine Darwinists had traditionally been concerned to defend asserted only that acquired characteristics could not be passed on down the germ line and were not heritable via meiosis and sexual reproduction. If previously imprinted methylation patterns were indeed not to be passed on in this way, then a complete demethylation (and subsequent remethylation) of the genomes involved had to be assumed, and precisely this did become for a time the most widely accepted account of what imprinting involved. But while empirical evidence of the occurrence of demethylation in particular cases was soon obtained, it cannot be demonstrated simply on the basis of such cases that demethylation will in all instances be comprehensive and complete, and it now seems as if hope has failed to triumph over experience here.

Over time, evidence has accumulated suggesting that acquired methylation patterns can persist across generations; it now appears to be generally accepted that inheritance of this kind can occur, and may possibly have occurred to a significant extent even in human populations. The relevant laboratory evidence has largely involved mice and seems to have convinced most researchers that the induced methylation of "genes for" some specific proteins is heritable. The evidence for its occurrence in human populations derives from the analysis of historical records and although often impressively robust, these data are typically of a kind that is more readily open to alternative interpretations. The best known of this material relates to the birth weight of the children and grandchildren of malnourished women whose pregnancies corresponded to a

_____

ever, a difficult notion to grasp formally with any precision, and it is all too easy to misunderstand and misuse. We discuss it in chapter 5.

period of famine during the Second World War, especially in the Dutch famine of 1944. Abnormally low average birth weights were recorded, both in the generation of children born to these women, and in the next generation, born to parents now prosperous and well nourished who might have been expected to produce babies of normal weight. The finding is consistent with radically reduced levels of nutrition causing the transcription of specific DNA sequences to be suppressed by heritable changes in methylation patterns (Vines, 1998). Another recent example concerns a robust connection between paternal smoking in early adolescence and obesity of sons and grandsons (but not female descendants) (Pembrey et al., 2005). One possible interpretation of this connection is that environmental insults at specific points in development may give rise to heritable epigenomic changes.

Whatever else, the appearance of this kind of work has opened a space for debate and controversy about some of the central collective commitments of biology. Earlier work is now being extensively trawled and potentially anomalous findings previously ignored or explained away are now being reinterpreted within the new framework. And as is typical in such episodes, a fascinating picture is emerging of how the dominant Darwinian/Mendelian frame had previously been used as a mandatory sense-making resource with which whatever findings researchers produced had somehow to be reconciled. At the same time, various rhetorical devices have been employed to polarize and intensify the controversy and magnify perceptions of its importance. Proponents of epigenetic inheritance have inflamed the debate by proclaiming themselves Lamarckians, provoked in some cases perhaps by the dismissive responses that initially greeted their ideas.[20] Some of them have happily explored the most speculative and far-reaching possibilities of their Lamarckian views and the ways in which they could entail radical revisions of existing accounts of evolution, and although the resulting debates have generally had a solid technical basis, they have inevitably also had a political dimension.[21] The maintenance of the authority of the theory

20 Jablonka and Lamb (2005) use the term, though they argue strongly against the interpretation that Lamarckism means direct feedback from the environment to hereditary material; West-Eberhard (2003) has been interpreted as presenting a neo-Lamarckian picture, though she seems more resistant to the application of this still generally derogatory term.

21 Given our earlier discussion of theories as cultural resources, we should expect that their use in evolutionary debates would not be confined to those with a given political orientation or set of existing commitments, and indeed the

of evolution is widely regarded as a sine qua non for the continued support and standing of professionalized biological science itself, and in some societies as an essential part of the defenses against profoundly anti-naturalistic, religiously inspired critics who have significant political support. And the fact that most of the enthusiasts for epigenetic inheritance themselves rely on naturalistic arguments and are in no sense opponents of science has not always made their challenge to this authority more acceptable.

Even though the enduring persistence of imprinted methylation patterns would constitute no real threat to the authority of evolutionary naturalism in biology, it would still be more convenient if they turned out in the main to have an easy-peel character. But it is not only in the context of debates about evolution that the adhesiveness of methylation patterns has this kind of nontechnical relevance. In the context of ontogeny, it has a still more direct and far-reaching moral and political salience. As we have just noted, the human body is composed of over a hundred different types of cells, and cells of different kinds are increasingly likely to have different ethical, moral, and legal statuses attached to them. But where different moral or ethical statuses are assigned to cell types on the basis of the natural differences between them, those differences are generally traceable to differences in the proteins they produce.[22] And these differences in their turn, along with the extent of their persistence and durability, appear to be the results of different methylation patterns and/or similar patterns of epigenomic variation.[23]

Historically, far and away the most important moral distinction between human cell types has been that made between somatic cells and cells in the germ line, particularly fertilized egg cells or zygotes, which have sometimes been assigned pro forma the same status as a fully formed human being.[24] Currently, however, attention has turned in another direction as intense debate has arisen about the ethical sta-

---

existence of imprinting has been used to advance evolutionary psychology and extend traditional selectionist views of human evolution (Haig 2002).

22 It is also possible, of course, for intrinsically indistinguishable cells to be accorded different moral and ethical statuses, on the basis of their developmental histories for example, or the settings in which they are encountered, or the functions they perform in those settings.

23 There may be other causes of cellular difference within the same individual, such as, for example the X chromosome inactivation process identified in chapter 2 as the cause of mosaicism in human females.

24 We discuss some of the issues raised by this in chapter 7.

tus of human embryonic stem cells. These cells are precursors of the many kinds of specialized cells making up the human body and are increasingly attracting the attention of researchers as promising therapeutic resources with potentially enormous benefits for human health. At the same time, however, powerful bodies of opinion have persistently questioned the ethics of such research. At first, the major reservation concerned the source of the cells, the harvesting of which entailed the disruption and destruction of human blastocysts or very early human embryos. But the debate has since broadened and now involves claims about the nature of the cells themselves, and increasingly about their intrinsic powers and potential. Many lines of these versatile cells appear to be capable of engendering all the specialized kinds of cell that make up the body of a human being, and any such cell, it is said, because it is potentially a human being, should be treated with the respect appropriate to a human being and not be exploited as an instrument of the purposes of others. In effect, because of their potential, these embryonic stem cells are themselves embryos, and as such (it is said) they are human beings and not to be violated.

This is not the place to discuss the deluge of casuistry that has poured over the issues here, or to ask what moves people to accept such accounts of human stem cells. All that is relevant here is that the putative difference between "totipotent" cells, from which complete human beings might potentially be produced, and "pluripotent" or "multipotent" stem cells of lesser potential, from which they supposedly cannot be, has become one of considerable importance. This is an epigenetic rather than a genetic difference, at least if the authority of current biological science on the issue is accepted.[25] But it is not only differences in epigenetic states that matter here: with more and more emphasis being placed on potential, the extent of the stability and the possibilities for change in the hypothesized epigenetic states are increasingly going to matter as well. In other words, the extent of the stability of methylation could turn out to have great ethical significance in this context.

25 Currently it seems to have been established empirically that stem cells are interestingly different epigenetically, and in methylation patterns in particular, from other cell types, even if how relevant the differences are to the functioning of the cells remains an open question. One interesting incidental finding of this research is that embryonic stem cells might be differently methylated or imprinted depending on the time of their harvesting. This would represent a possible technical obstacle to the reliable use of such cells.

Law, ethics, and morals are concerned with the ordering of human conduct. This makes them especially appreciative of schemas that divide the natural world into permanently constituted distinct and separate kinds of thing that need and require distinct and separate kinds of treatment. Part of the attraction of the old genetics, particularly in the versions familiar to nonspecialists, was that it seemed to offer a classification of phenotypes of this sort, one that would permit clear and simple rules of conduct to be built, for example, on supposedly natural and permanent differences in the natures and the powers of men and women. But of course, not even the old genetics provided this, as the case of sex, properly addressed, nicely illustrates.[26] And when we move from genetics and the kinds of organisms to epigenetics and the kinds of cells, we are in a different ballpark altogether.

One can imagine why ethicists, lawyers, and regulators might possibly wish to draw up special rules for the treatment of those cells with the potential to engender a complete human being, but they would be unwise to think of that potential as a fixed characteristic of cells themselves. Unspecialized cells are generally reckoned to be more potent than differentiated, specialized somatic cells because of epigenetic differences such as different distributions or levels of methylation. But the assumption that the methylation patterns in specialized cells significantly limit their potency implies that the patterns cannot be readily removed. Even if nothing counts against this assumption at the moment, it may turn out to be incorrect as more empirical studies are carried out,[27] and, more profoundly, it may be undermined by technical advances. There is nothing to say that we will not eventually be able to create environments wherein cells of hitherto limited potency reveal themselves to be totipotent after all.[28] Potency, it is worth bearing in mind, is context dependent

26 See Dupré (2002, ch. 8); and for details of the molecular genetics, Holme (2007).

27 Unsurprisingly, as more and more empirical research has been done, so estimates of the potency of given stem cell lines have tended to be revised upwards.

28 If we think of methylation as a kind of gluing, or even if we think in terms of the intrinsic strength of chemical bonds, we encourage ourselves to see the result as stable and hard to reverse. Perhaps instead we should explore the analogy with locks and keys. A lock may permanently close a space as far as most of us are concerned, yet open in a second to the possessor of the key. And similarly, it may be that methylation patterns of such great stability and such high observed heritability as to prompt talk of permanence will disappear in seconds when the appropriate key appears, whether in the shape of a natural enzyme or an ingenious scientific intervention.

and not wholly determined by factors internal to the object to which it is attributed.

It is a general feature of the biochemical processes that constitute cellular activity, as it is of all chemical reactions, that they are reversible. Of course, reversibility may be possible in principle but difficult in practice: putting the toothpaste back in the tube and reconstituting the acetone peroxide after the explosion are examples. But it is an interesting feature of cellular reactions that very frequently they proceed in both directions within the environment of the cell, whether at the same time or in different conditions and circumstances. This is evidently the case with methylation and other epigenetically significant processes, which presents a less than ideal scenario for those regulators and ethicists who would like to distinguish between clearly demarcated cell types with different powers and potentials and rest safe in the knowledge that one type of cell cannot be transformed into another. The natural world itself, of course, is wholly indifferent to the problems of ethicists and lawmakers and under no obligation to supply them with such cell types. But the researchers who work on this area of the natural world have been more considerate, and have been willing to characterize cells as if their supposedly different levels of potency were stable intrinsic characteristics.

It is of course precisely the versatility of stem cells that makes them foci of interest, and however problematic notions like totipotency may be, research on stem cells does require some way of referring to that versatility. But researchers can hardly fail to be aware that there is no "real" level of versatility or "potency" intrinsic to a stem cell, or indeed to any cell, and that what a cell has so far done is not a reliable indication of what it has the capacity to do.[29] Researchers are also well aware, however, that regulatory and ethical difficulties arise precisely from this same property of versatility and that where cells are identified as "totipotent," opposition to their use is more likely to be encountered. Hence, if scientists otherwise prone to hyperbole in describing the promise and potential of their work should occasionally err on the side of modesty when they describe their cell lines and characterize them as "pluri-" rather than "toti-" potent, we should not be too surprised. There is already a sense that this distinction is becoming unsustainable and may

29 Also relevant here is what cells are capable of with the active assistance of scientists. If a somatic cell could be transformed in the laboratory into a totipotent stem cell, would that mean that the former was totipotent, and if not, why not? The question is not purely academic: Dolly the sheep was famously cloned using the nucleus of a mammary cell.

now be little more than a convenient fiction (Hauskeller, 2005). Certainly, it should not be taken for granted that the less than totipotent stem cells studied by some groups of scientists will still merit that description in a few years time, or even that the classification system founded on extent of potency will still be viable then.

New similarities and new differences

As the focus of attention moves away from genomes to methylated DNA, and further on to transcribed RNAs, polypeptides, and proteins, life gets more and more complicated. As far as the present discussion is concerned, however, life is complicated enough already and we shall not make this move here. Even before our brief excursion into epigenetics, it will have become clear that the data produced by genomics and the new genetics allow us, if we wish, to document vast numbers of new similarities and differences among living things and to address the problem of ordering and sorting them at a new level of complexity. And this immediately brings us face to face with the completely general problem we now want to discuss, of how the empirical similarities and differences between things may be used to classify them and group them into a limited number of kinds.

The problem here is a profoundly difficult one because of the commonplace observation, which received its classic philosophical elaboration in the work of Nelson Goodman (1954), that between any two objects there are countless similarities and countless differences. Goodman's work was particularly influential through its reformulation of the traditional problem of induction. If we are to learn from experience, we need to identify cases in the present that are similar to cases we have encountered in the past. But given the limitless numbers of similarities and differences between the present and the past case, how can we ever be justified in applying past experience to the present in one way rather than another? Goodman's response to the question, like David Hume's before him, was ultimately skeptical. Certain ways of classifying things or situations are deeply entrenched in our linguistic practices, and their entrenchment provides us with the confidence to project them into the future even if it does not provide us with any justification for doing so. W. V. O. Quine gave a slightly less skeptical take on the problem by suggesting that our past classificatory tendencies had been refined by a process of natural selection: "Creatures inveterately wrong in their inductions have a pathetic but praiseworthy tendency to die before reproducing their kind" (1977, p. 126).

There is no solution to the problem of induction. And neither is there any solution to the problem of ascertaining how far two things are truly the same as each other empirically, or even truly the same in some given empirical characteristic. Neither Goodman nor Quine nor anyone else has been able to offer a solution to either problem. It is a measure of the greatness of David Hume as a philosopher that he never even aspired to offer a solution. Equally, it is a measure of our reluctance to accept that such problems are indeed insoluble that Hume's brilliant diagnosis of what makes the problem insoluble has so often been referred to as "Hume's solution" to it. However reluctantly, we simply have to accept that we could always turn out to be wrong the next time we make an inductive move, and that the next time we treat something as the same as another already-known thing, we could end up deciding that we should not have so treated it after all. We cannot discuss here the arguments that have long surrounded this conclusion, but we cannot ignore it either. We have to make mention of it, however dogmatically, because of its relevance to the discussion that follows. In a nutshell, it reminds us of the need always to question assertions that one thing is the same as another, mindful of the fact that it could equally have been asserted that the two things were different. And given what is to come, it is worth adding that this can also be said of assertions that one thing is "98% the same" as another. Assertions of sameness need always to be addressed with a certain skepticism, even if in due course we come to accept them as having pragmatic value.

In what follows we shall adopt a limited, pragmatic attitude to the problem of sameness rather than a fully skeptical one. It suffices for our present purposes to note that particular scientific projects always involve quite specific methods and strategies for classifying like with like, that the different classifications that ensue are always contestable and subject to change, and that those that exist at any given time are not always easily made compatible with one another. These are points that need constantly to be borne in mind in understanding and evaluating claims deriving from scientific projects. And because we are often inclined to treat (what we take to be) similar cases similarly, they also have vital practical implications, as we shall see.

Classical genetics involved the systematic study of manifest similarities and differences in phenotypes; its main theoretical achievement was to explain the differences between sets of (similar) phenotypes by reference to invisible factors transmitted from one generation to the next. But with the advent of genomics and the attendant technological advances, these factors have in effect been made visible, and observations of their characteristics are directly recorded as A's or C's or G's or T's in the output of an automated sequencing machine. Genomes can now be described as four-letter sequences billions of letters long, and this has made unimaginably large numbers of new relations of similarity and difference accessible to us. A developed awareness of the vast number of ways in which quite short sequences of these letters may be identified as similar or different was indeed essential in the assembly of whole genome sequences in the first place, when bits of genomes had to be identified as adjacent on the basis of overlaps of similar sequences at their ends. Assessments of similarities of much the same sort continue to be the basis of many current branches of genomics, including the field of evolutionary genomics to be discussed in the next chapter, and the methods developed to make them have diffused into other life sciences as tools and resources. Much of the rapidly growing field of bioinformatics is concerned with applying high-speed computers to the identification of features of genomes (qua nucleotide sequences) through which different genomes can be compared. And of course this power to classify genomes facilitates both what scientists do and what those in the societies that support their work do as well. Familiar examples of the latter are our use of DNA "fingerprinting" reliably to identify any specific individual in our population, and the use of rather similar techniques to establish ancestry and identify individuals as of the same kin group or having the same ethnic origin. These are all relatively simple techniques, employing only a tiny proportion of the information available in the DNA, and yet they can easily individuate many billions of organisms and establish innumerable links and relations of sameness between them. And all of the recognized similarities and differences alluded to here are no more than a minute proportion of those that could be fastened upon and exploited as a consequence of continuing research in genomics and cognate sciences.

There is indeed a whole new domain of opportunities here, provided by the vast amounts of empirical knowledge genomics is generating, but it is also a whole new domain within which skepticism and reserve are required as innumerable claims about the similarities and differences between things are put forth. An amusing but by no means trivial example of the need for care when confronted with such claims can be found in Steve Jones' book *The Language of the Genes* (1994). Both the extent of the genetic similarities among individual human beings and those between *Homo sapiens* and other species are discussed in this book. Speaking of the relations of individual humans with their close kin Jones notes how "non-identical or fraternal twins . . . have half their genes in common and are no more similar than brothers or sisters" (p. 235). But comparing humans in general with chimpanzees, he cites evidence that "humans share 98% of their genetic materials with chimps" (p. 37), and goes on to observe that "a chimp may share 98% of its genes with a human being but it is certainly not 98% human" (p. 38).

These are rough-and-ready statements not intended to be models of precision, but even so the extent of the clash when they are juxtaposed as above is very striking. They suggest, prima facie, that there is more genetic similarity between a given human being and a chimpanzee than there is between her and her brother or sister. This is, to say the least, surprising. What are we to make of it? Certainly, Jones is open to criticism for some sloppy writing, but he is not straightforwardly wrong in either case. There is no "true" or "real" level of similarity we can cite in order to show that he is incorrect. And it is easy enough to see what he had in mind when citing these apparently contradictory estimates. The alleged 50% of genes shared by human siblings is defensible as an allusion to genetic similarity produced specifically by descent, that is, by copying from parental genes. The 98% of genes allegedly shared by humans and chimps seems to be an allusion to the results of direct comparative studies of the totality of human and chimp DNA. Imagine lining up the human genome against the chimp genome and finding that they match or correspond along 98% of their length.[30] These two notions of "sharing" are quite different, but they are hopelessly confused in most of the literature available to nonspecialist readers, which serves more to mislead than to enlighten.

The confusion that results from conflation of these notions of sharing and the associated measures of similarity is exemplified by the ubiquitous myth that siblings have only half their genes in common, a myth propagated in text after text by popular writers. No doubt part of the reason that this misleading claim is so widely accepted is that an account that is both accessible and accurate is hard to provide. A difficult distinction has to be drawn between two ways in which the genes of different organisms may be said to be "the same," one referring to sameness of nature or constitution, the other only to sameness that is

---

30 It is tempting to think of a percentage of matching base pairs in the two sequenced genomes, but this too would vary with method of measurement, and in any case Jones did not have such estimates available. The method of matching he probably had in mind was the DNA-DNA hybridization technique he describes on p. 129, which led to highly controversial results when applied to human and chimp DNA.

the immediate consequence of inherited transmission. Even Richard Dawkins with all his renowned expository talents explains these differences properly only in the footnotes of *The Selfish Gene*. But it is also important to note how nicely the mythological account chimes with everyday intuitions, according to which it seems natural to individuals "already sophisticated in the intuitive calculus of blood ties" (Wilson, 1975, p. 118) to take sanguinary closeness as a measure of the proportion of the genes two individuals have in common, with sibs, or parents and children, having half the same, grandparents and grandchildren a quarter, and unrelated individuals none at all. Busy and/or lazy writers have been tempted to exploit this erroneous way of thinking and hang their expositions upon it. The results can be close to gobbledegook, as when Stephen Pinker (2002, p. 374) invites us to compare "biological siblings, who share half their genes and most of their environment, with adoptive siblings, who share none of their genes (among those that vary) and most of their environment."

It is useful here to contrast the situation of insiders, in this case researchers, and outsiders like the intended readers of this book, as well as its authors. Insiders alone have access to the situation of use of their own knowledge and training in the methods employed to produce it, and to that extent they are better equipped to grasp both its value and its limitations. Outsiders, lacking both in access and training, are far more vulnerable to accounts that gloss over these limitations and, for whatever reasons, misrepresent what is actually known. It is particularly striking how frequently the methods by which knowledge is acquired are omitted from popularizations of science. This deprives the reader of proper insight into both the value and the limitations of scientific findings. Writers can then make free with descriptions that have specific roles and functions in the contexts they are drawn from but that, when divorced from them, may be little more than a repertoire of abstractions for selective use in storytelling. However, in order to avoid committing this same sin of excessive abstraction ourselves, we need to provide a more specific illustration of the problems that can arise from it.

Let us begin with a very simple toy example. Think of two sets of very tiny solid rods, known to the mind of God perhaps, but not to us, to be rods that differ only in one place; perhaps there is a tiny hole or bubble at one end of the rods in the one set but not of those in the other (Fig. 1A). At the outset we know only that these rods are different— 100% different we might say, as we think of them simply as belonging to two distinct kinds. But we would like a more refined view of "how different" the materials of the rods are. We invent a method. First we

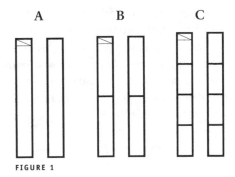

FIGURE 1

separate the rods en masse, even though we can't see them individually: one kind of rod floats in water perhaps and the other kind sinks, producing two different kinds of sludge. We feed the two sludges through a crushing and grinding machine capable of breaking each of the rods into two pieces. Use of this, and further separation and comparison of the products might incline us to say that half the materials in the two sorts of rod are similar, and half different (Fig. 1B). Were we to improve our crusher so that it broke all the rods into four pieces, we might conclude that the rod materials were 75% similar and 25% different (Fig. 1C). And as our methods changed in the future, so our estimates of likeness/difference would change. Not being God, we would never have perfect knowledge of what our methods were "really doing" to the rods, and neither would we ever arrive at a method-independent way of identifying the "real" degree of likeness between the two kinds. Even so, we might perhaps alight upon a method we found adequate for all existing practical purposes, and by being aware of the method dependence of our estimates of likeness, we should know to be cautious in relying on those estimates for different purposes.

Any similarity between the long thin rods just discussed and current notions of DNA molecules is of course wholly intentional. And indeed comparisons of the DNA of different species used quite commonly to be made using methods that manipulated DNA physically and chemically as tangible material and compared the effects of doing so. Methods of broadly this sort have been the underlay for claims like that of Jones that "humans share 98% of their genetic materials with chimps," as we noted earlier. The problem here is that there are very many ways of characterizing the genetic similarities between the two species, and indeed many different ways of assessing the similarities of their "genetic material," and they yield different estimates. If Jones' assertion conveys the idea that humans and chimps are very alike in a scientifically meaningful sense, it is because of the authority of the writer and our willingness to trust someone of his standing to pick an appropriate figure from the available possibilities—in this case a figure that dramatizes the similarity of chimps and humans that is currently part of the accepted wisdom of science and makes it memorable. At the present time, of course, the kind of similarity at issue here is generally studied by comparing four-letter genome sequences stored in computers. But again there are innumerable methods of making such comparisons, each liable to yield a different result, and the choice of method by researchers remains dependent on their aims and purposes, as well as on what technical and instrumental resources they happen to have to hand.[31]

It is worth mentioning as well that even in an informatic frame we are not compelled only to match A with A and C with C. As discussed in the previous section, it is only a reflection of particular theoretical commitments that we count 5-methyl-cytosine as a form of cytosine. If we chose to regard it as a different base, DNA sequences would seem less alike and a whole range of existing similarity claims would be weakened. On the other hand, we could move in the opposite direction by noting that cytosine is very similar to thymine and guanine to adenine, and comparing DNA strands as two-letter purine/pyramidine sequences. And from the point of view of a doggedly reductionist physicist, any two DNA strands could be said to differ only in the odd proton or neutron here and there, making us 99.99% similar to chimps perhaps, and 99.9% to bread mold. Whether there would be any point or purpose in working out these degrees of similarity and difference is another question entirely.

To summarize: in the context of scientific research, assertions of likeness and difference are usually the products of thoroughly familiar methods, widely known in terms of hands-on experience as well as through verbal descriptions, and the same is true for the most part in the sphere of technical applications. For specialists, a vast array of practice and activity constitutes the context in which their discourse functions. It makes the import, the utility, and the limitations of their verbal formulations of likeness/difference clearer to them than to outsiders, and makes the latter more vulnerable to misinterpretations and misrepresentations of those formulations, whether deliberate or accidental. Nonetheless, it has to be recognized that scientists' judgments of sameness and differ-

---

31 The case in point, and the difficulties in arriving at specific figures for such comparisons, is well discussed by Marks (2002).

ence are frequently interesting in more than a narrowly technical sense, and may be very widely used as the basis or the justification of decisions to treat objects or organisms in similar or different ways. Think again of the 98% similarity of human and chimp genetic material mentioned by Steve Jones. In fact, science seems now to be suggesting an even higher degree of similarity. Comparing only the coding DNA of the two species, we find 99.4% similarity, or so we are now told (Wildman et al., 2003).[32] And this finding is now being cited in efforts to make us more kindly disposed to chimps as a species.

It remains wholly unclear why we should discriminate morally in favor of chimps solely and simply on the basis of their DNA, but the example does serve nicely to illustrate how a relation of sameness of a highly specific sort and a highly contextually determined relevance can be abstracted from its appropriate technical-scientific context of use and then redeployed as if it were a morally salient similarity. Or rather, it serves nicely but for one slightly problematic feature. Scientists themselves have been involved in carrying this particular finding from the technical context into the context of moral debate. According to Morris Goodman, it supports "the idea that it wouldn't be ethical to treat [chimps] in the way laboratory animals like rats or mice are treated" (*The Independent*, 26 May 2003, p. 6). Evidently, scientists themselves may sometimes be tempted to further broader agendas by detaching their technical assignations of similarity and difference from the esoteric practices and purposes that give them their semantic stability.

Reducing complexity

The problems discussed in the previous section started off from the uncontroversial point that between any two physical objects there are countless similarities and countless differences. Another way of making this point is to speak of the complexity of the physical world: complexity is an idea very prominent in the current scientific *Zeitgeist*, and if our previous starting point is expressed in terms of this idea, a new direction is opened to the reflections prompted by it. A cell, we might say, is indefinitely complex; however complex our models and representations of it

32 Though even more recently, a much lower figure has been suggested, of 94%, based on comparison of the extent of duplication of genes (Minkel, 2006). The growing diversity of such estimates, of course, emphasizes the absence of any uniquely and objectively correct answer to this question, and makes it plausible, at least, that other concerns may sometimes be involved.

are, there will be further features of it that are left out of account. There is no complete description of a single cell, let alone of the whole world. As our earlier discussion tried to illustrate, all descriptions of things simplify the relations between the things. And science, far from working to reduce the gap here by producing ever more detailed descriptions, is constantly looking for useful ways of simplifying its descriptions of phenomena. Indeed, in science as elsewhere simplification does not begin with description; even perception simplifies, through the filtering effects of the perceptual apparatus used and through selective attention to just some aspects of the information it provides. Verbal descriptions involve a further level of simplification beyond this; they must deploy words that refer indifferently to a range of more or less different things, and hence any given description will inevitably ignore some differences between the things to which it refers. Moreover, if we turn to the sciences and the highly theoretical language they employ in their descriptions of things, we find even greater selectivity and still more that is ignored.

Contrary to first impressions perhaps, bioinformatics is a theoretical/modeling enterprise that serves to reduce the information we have to deal with. The vast four-letter arrays of genome sequences represent a massive reduction not merely of the information in principle derivable from genomes but of that which is already known and codified. A string of A's, C's, G's, and T's tells us nothing at all about shape or geometrical configuration or indeed about most of the features of the components of the genomic DNA. And as we have explained, even the nucleotides supposedly described as A's, C's, G's, and T's are themselves abstractions from a larger set of nucleotides actually found in "real" genomes. Of course, one reason that bioinformatic models, like all scientific models, provide an insufficient basis on which to make empirical predictions about real-world states of affairs, in this case about how actual pieces of genomic DNA will behave in situ, is that all this information, and more, is missing. And indeed it is a fascinating philosophical problem to understand how it is that such models have any predictive value at all, as evidently they do.

Despite these massive simplifications, however, the sheer quantity of information that genomics is now able to provide about the molecular systems in particular cells or organisms conveys an inescapable sense of their complexity and prompts the question of how the functioning of such complex systems can possibly be understood. And at another level many years of data gathering are creating a strong sense of information overload in those who reflect on the trillions of nucleotides now stored as sequences in genomic data banks and the comparable volumes of data on RNA transcripts and protein products now augmenting them. The thought is that there is simply too much information here to process as it stands and that selective attention to the most relevant and important data would greatly simplify the task of making use of it and drawing interesting conclusions from it. Thus, at both levels an urgent priority has come to be that of reducing complexity: the question has become how to transform what first present themselves as highly complex problems into simpler, soluble problems, while ensuring that the solutions to the latter continue to tell us something valid and useful about the world.

Attempts to effect radical reductions in complexity figure large in a number of current research programs and are often explicitly cited as a part of their rationale. An important instance is synthetic biology (Bedau, 2003), which is devoted to the laboratory synthesis of complex biological molecules and other biological structures. It has been described as a supercharged version of genetic engineering, something toward which genetic engineering as we now know it was the first faltering step, and its progress may be measured in terms of the size and complexity of the biological objects it proves capable of synthesizing. But at some point, which point precisely being a matter of how these things are defined, the most complex synthesized object is also going to be the simplest extant organism. This at least is the view of some synthetic biologists, who now see the production of a simple viable organism as the goal definitive of their field and the touchstone of its future success. On the face of it good progress has been made toward this goal. Laboratory-synthesized DNA of whatever sequence is desired is now readily available, and its production cost in dollars per base is declining very rapidly. Biobricks, in which several DNA sequences are combined with other molecules in modules that perform a single function, are now becoming available as components with which to build larger biological systems. The first virus was synthesized some years ago now, and others, sequence identical to naturally occurring varieties, have been created since, although unlike the bacteria on which many of them prey and whose generally far longer genomes they exploit, viruses are not everywhere counted as living organisms. Recently, however, a successful transplant was announced, wherein a bacterial genome was inserted into a cell of another kind of bacterium and took over its functions, suggesting that the practical problem of synthesizing a new bacterium now largely amounts to the problem of synthesizing a bacterial genome,[33]

33 A solution of this practical problem is not in itself a solution to the problem of creating life from inert matter, in that the membrane and cytoplasm of an

and firm plans have indeed already been announced for the synthesis of a simple synthetic bacterium, *Mycoplasma laboratorium*.[34]

What we have here, it might be thought, is an effort to minimize complexity emerging as an incidental aim of a great project to synthesize life itself. But in fact things seem to be the other way round. Synthetic biologists are aware that many people are likely to attach a profound significance to their project, and that many ethical perspectives, especially religious ones, identify the project of creating life as sacrilegious or deeply offensive to the dignity of existing life. They are very far from being oblivious of the arguments that could be directed against them by those with perspectives of this sort, the kinds of argument we discuss at some length later, in chapter 7, but they seem little troubled by them. A patent application has just been made in respect to *Mycoplasma laboratorium*, and possible methods of making and using it, but as we write it is not expected that the patent will be granted.[35] It is tempting to suggest here that the application might also be understood as a symbolic expression of a materialist and/or instrumental attitude to "life itself," and of the lack of any reserve about its future manufacture and commodification, at least in its bacterial manifestations.

There seems to be relatively little interest in the context of synthetic biology in simply showing that life can be created from nonliving matter, or in reflecting on where precisely the boundaries may lie between what is alive and what is not. The project appears mainly to be driven by other imperatives. The synthesis of a uniquely simple organism, ideally the organism with the simplest genome compatible with viability, will, it is hoped, facilitate the study of regulatory pathways and switching mechanisms in the simplest available system and facilitate a quite

---

existing bacterium are involved, but as we go on to note the latter problem is not the major concern of synthetic biologists, any more than the achievement of true cloning was the primary concern of the creators of Dolly the sheep.

34 These recent developments have involved the Venter Institute, founded by J. Craig Venter, one of the most eminent and interesting figures in the history of genomics, whose individual career marks out many of the most important developments in the field. One of the most innovative and dynamic contributors to genome-sequencing methods and notably to the sequencing of the human genome, Venter then made a seminal contribution to environmental genomics (Venter, 2007, ch. 4.3) before moving into synthetic biology.

35 Again, the application is from the Venter Institute. One of the contemplated uses of this synthetic organism is in the creation of microbes to combat climate change by serving both as fuel sources and as fixers of carbon. Application no. 20070122826: www.uspto.gov.

general understanding of how its components interact. And this, as well as being sought after for its own sake, is seen as opening the way to a range of economic applications. The thought is that a "stripped-down" genome will serve as an improved chassis to which additional special-purpose DNA can be added to create different organisms suited to a range of tasks. And indeed these tasks are by no means minor ones. One current project already close to completion stands out here as both of vast potential importance economically and capable of serving as a paradigm for similar projects in the future. Teams of researchers are using both *E. coli* and yeast, the standard workhorses in this context, to create organisms that will radically lower the cost of producing artemisinin, an effective and increasingly important treatment for malaria, currently the cause of over a million deaths a year, predominantly in the world's poorest countries (Keasling, 2007).

Another program of major importance in which the reduction of complexity is a predominant concern is integrative or systems biology. This can be seen as a further development of mathematical computer-based (in silico) methods used in genomics and, like synthetic biology, is partly the product of former genomics researchers carrying their methods into new contexts. Centers and departments with this title are currently being established in most parts of the developed world, and making ambitious claims as to what they hope to accomplish. Systems biologists recognize an urgent need for methods of reducing information overload, even though they also argue that without the huge accumulation of biological information that has occurred over the last twenty years their project would not be possible. The aim of the project is to use powerful information technologies to make inventories of the more important molecules in biological systems (DNA, RNA, and proteins, in particular), and to develop modeling techniques that provide selective yet accurate representations of the biological processes in which they are implicated. With ever better in silico models of biological subsystems, it is hoped to explore the effects of introducing perturbations into them. A strong economic-utilitarian impetus to the project is apparent here: it is hoped, for example, to investigate the effects of candidate drugs without invading the bodies of either human or animal test subjects.

There are some very different views as to how systems biology should proceed.[36] One broad division is between "bottom-up" and "top-down"

36 We can say little about the range of specific systems the field aspires to model. Ideally, closed molecular systems are required, but there are none in biology. There are, however, systems that can be regarded as closed for given

approaches. Bottom-up systems biologists think that with a sufficiently comprehensive molecular inventory (a "parts list") one should be able to work up from the bottom and gradually produce a model of a biological system sufficient to allow predictions of its response to environmental events. However, as models are always simplifications, modeling strategies require some rationale for deciding what can be left out and what must be included. Generally, bottom-up systems biologists advocate some kind of iterative methodology here: research will move back and forth between wet experiments and computer simulations until enough of the important factors are incorporated into the model to produce sufficiently reliable empirical predictions. In contrast, top-down systems biologists doubt that the problem can be solved in this way and argue that it is necessary to begin with a fairly rich theoretical framework so that attention is focused on the relevant aspects of the system from the start. Of course both these perspectives are somewhat caricatured here, and actual projects often include both bottom-up and top-down elements.[37]

A central problem for all versions of systems biology is how the quantity of information used can be reduced to a manageable level without thereby rendering its models useless. Here the bottom-up approach at least offers a coherent inductive strategy: keep checking back with empirical findings and adjusting the model, hoping that it will not be long before it becomes adequately predictive. But of course there is no guarantee that the model will ever turn out that way, just as there is no guarantee if one tries to proceed on the basis of general principles. This is not a criticism of systems biology: all attempts to model empirical states of affairs involve strategies of complexity reduction that may or

---

purposes and others that are near enough closed in normal circumstances. The most clear-cut instances of partially isolated molecular systems in biology are cells, and models of whole cells are sometimes identified as ultimate goals of systems biology analysis. But so far ambitions are largely restricted to specific biochemical networks, and conversely, it remains legitimate to ask of substructures of cells, or larger sets of cells—organisms, or perhaps microbial communities—which of these are sufficiently isolated to be worth modeling as separate biological systems.

37 The first of these tendencies is often associated on with Leroy Hood, director of the world's largest institute devoted to systems biology, the Institute for Systems Biology in Seattle, and the second with Hiroaki Kitano, the leading Japanese systems biologist; in each case, though, the degree of caricature mentioned in the text should be borne in mind. A more detailed discussion of this divide can be found in O'Malley and Dupré (2005).

may not turn out well. And if there are grounds for more specific reservations about the models of systems biology, they result not from efforts to simplify but from the sheer scale and complexity of the models notwithstanding these efforts. These are models that invite comparison with those used in ecology or meteorology, which are also virtually constituted in silico and like those of systems biology exploit to the limits the resources of the largest computers.

Whether this analogy should make us more or less sanguine about the eventual success of systems biology is moot, although we should take care not to be led by tradition into underestimating the ability of meteorologists to predict the weather. With systems biology, and indeed most of the efforts currently being made to build upon and exploit existing achievements in genomics and molecular genetics, it is a matter of waiting on events. If and when scientists themselves become convinced that new methods have proved themselves and new achievements have been made, research will get behind them and seek to exploit them in their turn: this will be the sign of their success.[38]

38 Thomas Kuhn (1970) famously suggested that the crucial factor that allows a new field of scientific investigation to develop successfully is a specific explanatory or technical achievement that, apart from being impressive in its own right, suggests possible extensions to a range of new cases and can serve accordingly as a paradigm for future research. It remains a matter of controversy whether anything of this sort yet stands to the credit of systems biology, but work by Trey Ideker and colleagues (2001) on galactose utilization networks in yeast is currently seen by some as approaching paradigmatic status, and the authors describe the paper reporting this work and its findings as offering a "proof of principle" for systems biology methods.

4 Genomics and Evolution

Classification

One of the most important contexts in which genomics and genomic technologies are currently having an impact is that of the other biological sciences, where they are not only inducing changes in theoretical understanding but also contributing in a more diffuse way to what has been called the molecularization of these sciences. In this chapter we focus upon how they are affecting what is arguably the most important body of theory in biology, that concerning biological evolution. The theory of evolution is, of course, of great interest and importance beyond the narrow confines of the sciences. Variants of the theory are now established as a part of culture generally in many societies, and whatever affects the scientific understanding of evolution is liable also to affect the wider culture of these societies. But while we are well aware of links that could be made between what we have written here and this wider context, our explicit concern will be with the impact of genomics on biological theory.

Consistent with this, we shall approach the theory of evolution by way of a central technical concern of biology, and one increasingly tightly linked with evolution over the last century, its contribution to the task of biological classification. This will prepare the ground for a subsequent discussion of evolution in which we stress the relative neglect, in most existing accounts, of the evolution of molecules, notably of genomes, and also of the microbes that have long been and remain still the dominant life forms of the planet. We emphasize that the evolutionary history of these entities extends over a far greater proportion of evolutionary time than the more familiar evolutionary history of complex organisms and differs in a number of interesting ways from that familiar story. And we point out how reflection on the evolution of molecules and of microbes, increasingly informed by genomic research, could lead both to changed understandings of evolutionary processes, and even, as our concluding section on metagenomics suggests, to new ways of defining and identifying the individual biological organisms involved in them.

We begin, however, with a glance at the problem of biological classification. In the preceding chapter we described some of the resources genomics is providing for identifying things as the same or different, which are precisely the resources needed for classification. Predominantly, of course, these resources build on existing classifications and add to existing techniques, offering new ways of identifying similarities and differences. Although any two objects can be seen as similar and different in countless respects, for the most part there is just one taxonomy of living things currently accepted by biologists, and just a very few characteristics and relations are involved in creating it.[1] Why is this? One obvious answer is simply pragmatic: a single shared system is highly desirable for the reliable communication and storage of biological knowledge. But there is also a long tradition of supposing that the biological kinds that are really there can be represented in a single unique system of classification. Such a system used to be called a "natural classification," and some similarities were emphasized at the expense of others because they were thought most likely to lead us to a classification of the real, natural kinds.

This particular form of realism is not compatible with our own view that the dominance of one or another system of classification at any given time or place is always a contingent matter in need of historical investigation. But it is true nonetheless that some similarities and differences are generally treated as much more important than others, and that this selectivity informs biological taxonomy at every level. Thus, for example,

---

1 The paradigmatic area for classification is biological taxonomy, the project of distinguishing kinds of organism. Whereas "biological taxonomy" generally refers to the activity of classifying organisms, the science that aims to survey the diversity of life through time as well as space is known as systematics.

we currently tend to take it for granted that the developmental trajectory of an organism is something fundamental to it, and that its descent from a parent organism is an indication of sameness between parent and offspring so important that it identifies parents and progeny as things of the same kind. Analogously, with entire populations of creatures, believing as we do that living things are only ever produced by other living things and that the organisms produced are similar in a profound sense to those that produced them, we see relations of descent as fundamental to the project of biological classification. In fact this *phylogenetic* approach to taxonomy, grounded in relations of descent, has emerged over the last few decades as the dominant view of the subject, and it is as a tool for use in tracing *phylogeny* that genomics is making its main impact in this context.[2]

Many kinds of genomic similarities between organisms are now accepted as being due to relations of descent and are increasingly used in identifying and mapping those relations. Some of the best-known involve human ancestry. A striking recent case of the tracing of group origins involved the Lemba people of Southern Africa. The Lemba have a tradition that their ancestors were one of the lost tribes of Israel, and they have a number of practices such as circumcision, keeping one day a week holy, and avoiding eating pigs or pig-like animals (such as the hippopotamus), practices that are reminiscent of Jewish customs. Their case was studied by the British anthropologist, Tudor Parfitt, and a team of genetic anthropologists from University College London, who explored the origin of the Lemba from a genomic point of view. They discovered that in Y chromosomes of the men in a subclan of the Lemba there was a high frequency of a particular genomic feature, the so-called "Cohen modal haplotype," also characteristic of the Jewish priesthood in undisputed Jewish populations. This provided evidence that the Lemba might indeed have had a Jewish origin as they traditionally claimed.[3]

2 Even so, references to visible morphological characteristics—the traditional bases of biological classification—remain not just important but essential to phylogenetic classification. A very clear account of the general problem of phylogenetic inference can be found in Sober (1988).

3 The case is well known since it is more than a mere illustration of genomics having utility in the investigation of relations of descent. It is one of a number of cases where evidence of ancestry, as well as helping to confirm or refute a claimed collective identity, was also potentially material to claims for political rights and favorable treatment by administrators and bureaucrats (Hauskeller, 2004). We should stress that we do not offer a view as to what, if anything, this haplotyping evidence really does establish.

The actual evidence provided by genomics in this example, if not its interpretation, is less open to criticism than some for two main reasons. First, it derives from the examination of a chromosome, the single male Y chromosome, most of which has no counterpart with which to exchange material during meiosis. Unlike other human nuclear DNA, this chromosome is passed on to descendants largely unchanged through recombination with a counterpart derived from the second parent, and has proved especially interesting in that it permits the study of relations of descent confined entirely to the male line. The DNA of the mitochondrial genomes in the cytoplasm has a similar importance in that it is inherited entirely from the mother and facilitates studies of descent confined entirely to the female line.[4] Second, the time span over which these examples extend is so short that relatively few problems arise in identifying genome sequences that are similar due to descent.

Unfortunately, evolutionary phylogeny is mainly concerned with relations of descent that extend over periods of time that are orders of magnitude greater than those in the above example, and over such periods the problems of establishing similarity due to descent are very much more serious, whether traditional (morphological) methods are used or the newly available molecular methods of comparative genomics. Even the measurement of the time spans themselves remains a source of difficulty. Today "molecular clocks" are widely used to make such measurements: the number of point mutations accumulated along a length of genomic DNA is taken to be a measure of elapsed evolutionary time.[5] But even though molecular clocks are widely employed and have produced results recognized to be of major importance, a plethora of problems attends their use (Woese, 1987). Although many of these problems are narrowly technical, others raise deeper issues. In particular, in order to measure evolutionary time, molecular clocks need to tick with random mutations, but the clocks are actually a part of what is evolving and evolution is reckoned to involve the nonrandom process of natural

---

4 In both cases the assumption of sex-specificity has been questioned. It is hard to be sure that no cytoplasm, and hence possibly mitochondria, passes into the egg with the ingress of a sperm, and the Y chromosome does have an "autosomal region" that recombines with a part of an X chromosome during meiosis. Adequate allowance for these complications can probably be made, however.

5 The classic statement of the idea is Zuckerkandl and Pauling (1965). That point mutations will both occur and accumulate in a clock-like way is a corollary of the "neutral theory" of molecular evolution that we discuss in the next section.

selection. On the face of it, the theory that structures and informs the use of molecular clocks also gives grounds for distrust of them; and indeed, there would seem to be an intractable formal problem here, and it has certainly sustained continuing controversy.[6]

Needless to say, genomics has been able to contribute far more to phylogeny than methods of measuring and comparing vast periods of time, but to gain a sense of its greater contribution we first need to look at the objectives of traditional phylogeny. In brief, these have been to classify and distinguish the different kinds of living things—in practice the various species recognized in biology[7]—in terms of relationships of descent. It has become standard to depict species using a tree conceived as representing, vertically, a dimension of time. Successive slices through any branch of a tree are taken to represent populations of organisms related by descent, and the branching points in the tree are taken to represent speciation events, points in time at which one species has diverged into two. The project of taxonomy is sometimes thought of as embodying the goal of giving a distinct name to every branch of the tree. The smallest twigs will be the species, and larger branches will generally be groups, or *taxa*, of higher level. This conception is the central idea of the school of taxonomy known as *cladistics*.

The tree diagram itself symbolizes a key assumption of traditional phylogeny: when a species has differentiated, it does not recombine with

6 One pragmatic solution to the formal problem has been to use nonfunctional DNA sequences as molecular clocks, on the grounds that as they have no function selection will not act upon them. Unfortunately, these sequences are generally of no use over very long periods of time as the quantity of accumulated mutation eventually swamps the ability to compare sequences. For such periods, therefore, it has been necessary to employ sequences that are very highly conserved, assumed to be sequences that have essential functions for the organism. One such sequence is the 16S ribosomal RNA sequence used by Woese in his work on the Archaea shortly to be discussed. This sequence produces a subunit of the ribosome, the structure that hosts the translation of messenger RNAs to proteins. Recognizable versions of the sequence are found in all organisms, and show only very minor variations even within the largest groups of organisms. Evidently, this DNA sequence has been subject to very little selective change, and the small quantity of change that is observable is assumed to have occurred in a random, and hence a clock-like, way.

7 Species are sometimes subdivided into subspecies, varieties, and such like, and the relations between these can also be explored. It is generally taken as something like a matter of definition that species are the smallest theoretically significant group, however.

other species and can transform itself only by further speciating events: branches of the tree can lead only to smaller branches and cannot rejoin. Every point in the tree diagram, or more precisely every crosscut of a branch or twig, represents a population of creatures all descended from some ancestral group and all cut off from reproductive relations with creatures beyond the group. Such a population is called a monophyletic group. The descendants of such a population will of course share the ancestry of the population and become themselves members of the same group. Thus, when any branch of an ideally formed phylogenetic tree is sawn through, at whatever point, a monophyletic group falls off.

We need to recall at this point that the relationships of descent recorded in a phylogenetic tree are by no means records of observations. All that is available for examination is living creatures and their remains and traces, along with their many and various empirical characteristics. Somehow patterns of empirical similarity and difference must be identified; those that indicate relations of descent must be distinguished from those that do not; and phylogenetic trees must be painstakingly constructed from the former. The goal is to distinguish so-called *homologies*, features that organisms share as a result of their common descent, from mere resemblances. These, in turn, are employed to address the basic problem of phylogeny, one that recurs again and again as work proceeds, that of investigating three groups of organisms, A, B, and C, and deciding somehow or other which two are monophyletic relative to the third.[8] In less precise language, the problem is to decide which two are most closely linked phylogenetically, or, less precisely still, which are most closely related by descent. In practice these decisions tend to concern the relationships between what we antecedently take to be closely related groups— for example, humans, chimpanzees, and orang-utans—but in principle A, B, and C could be any three groups. In making such decisions, it is not just intuitions of extent of similarity and difference that are relevant or even ways of distinguishing homologies from mere resemblances. It is necessary to judge which homologies relate to *ancestral* properties and which to *derived* properties. A property is ancestral if it was possessed by an ancestor that existed prior to any of the groups being considered, and derived if it first appeared after the emergence of at least the earliest

8 Note again that this methodology presupposes that we already have some way of distinguishing the groups we want to relate to one another. Once we have them the hope is that iterated comparisons of sets of three groups will allow the construction of a phylogenetic tree incorporating them all.

of those groups (see Fig. 2. p. 121). It is derived properties that are crucial for phylogenetic inference, but again it is often difficult to say, and always more a matter of inference than observation, which properties are ancestral and which derived. Thus, the fact that both lizards and cows have four limbs is not currently regarded as good evidence that they are part of a monophyletic group that excludes snakes, since four-leggedness is believed to be a property of common ancestors of all of these groups. Snakes are thought to have lost this feature, but the crucial question is when. As a matter of fact biologists are confident that they lost it long after a group containing both snakes and lizards diverged from that which led eventually to cows. Scales, on the other hand, are regarded as providing good evidence for placing snakes and lizards in a group distinct from cows since no ancestor of cows is believed to have possessed them: within this taxonomic context scales are derived properties.

How might features of genomes be relevant to this kind of enquiry? Just as with traditional phylogeny it will not be merely similar features of different genomes that will be potentially useful but features believed to be (at least) homologous. Where sequences are believed to be homologous, they count as evidence of a relation of descent; where there is mere similarity and no grounds for believing that one sequence is a later form of another, the sequences may have independent origins.[9] Distinguishing homologous genome sequences from sequences that have arisen independently is not a trivial matter. Thus, there are many instances where distinct enzymes catalyze the same biochemical reactions, and in many of these the enzymes involved have significant sequence and structural similarity. There is no simple method of identifying homology in such cases. For example it was discovered as early as 1943 that a particular enzyme, fructose 1,6-bisphosphate aldolase, existed in different forms in yeast and rabbit muscle (Warburg and Christian, 1943). These two enzymes were long believed to be associated with different phylogenetic lineages, since their catalytic mechanisms were different and little structural similarity was evident. More recently, however, it has been shown that these enzymes have a similar crucial fold and structurally similar

---

9 To illustrate the distinction in terms of a classic morphological example, the bone structure of the bird's wing and the whale's flipper are considered homologous, deriving from a bone structure common to all vertebrate animals and inherited from a taxon ancestral to all vertebrate animals. On the other hand, the wings of the bird and the bat, qua wings, are not homologous because (we believe) the nearest common ancestor of birds and bats did not have wings: they evolved independently in each lineage.

active centers, suggesting that they may have a common ancestor after all (Cooper et al., 1996; see also Galperin et al., 1998).

Because of the difficulties in identifying homology, the concurrence of different methods, and particularly methods as different as those of genomics and traditional phylogenetics, can have a special evidential value. Most questions about the phylogenetic relations among mammals are now reckoned to have been resolved by the combination of these two approaches. There remains, however, an element of residual controversy even about human evolutionary origins. The overwhelming orthodoxy is that our closest relatives are the chimpanzees, from whom our lineage is thought to have diverged about five million years ago. However, one respected anthropologist, Jeffrey Schwartz, has been arguing for many years that we are actually more closely related to the orang-utan. Schwartz (2005) claims that a sophisticated morphological analysis discloses a much larger number of derived characteristics shared by humans and orang-utans than by humans and chimpanzees. The evidence from genomic analysis may point in the other direction, but Schwartz argues that it is only because of the great prestige of molecular methods that this has been taken to settle the issue. We have no intention of offering an opinion on this highly technical debate, but insofar as traditional morphology is viewed as useful corroboration of genomic analysis when the two agree, it should presumably be given some weight when there is disagreement. Whatever the fate of Schwartz's claims, therefore, they can be heard as a reminder that molecular arguments should not automatically trump more conventional ways of addressing biological questions.

While the problem of distinguishing homology from mere similarity arises everywhere in phylogeny, more specific problems arise when genomic methods are employed because genomic phylogeny does not necessarily coincide with organism phylogeny. These problems are worth brief attention here because they are important to much of our later discussion. We need to be aware that genome sequences can be homologous in three distinct ways, and they are referred to as orthologues, paralogues, or xenologues according to how they originate. When genomic elements are related to one another as common descendants from elements found in an ancestral species, the significant relation in general phylogeny, they are known as orthologues. A genomic element is said to be a paralogue of another element, on the other hand, when the one has originated from duplication of the other within the same genome. Finally, two genomic elements are said to be xenologues when their similarity arises from the transfer of an element between different taxa. Paralogues are known to be common in almost all organisms, and xenologues, as will be discussed in more detail below, are common in at least many. But unlike orthologues, neither paralogues nor xenologues necessarily tell us anything about the relations of descent between *organisms*, and serious errors are liable to arise if they are mistaken for orthologues and used as indicators of such relations.

Armed with these distinctions we can now turn to one of the most striking examples of the successful application of genomics in the realm of taxonomy, one that serves to illustrate both the remarkable insights that have been gained through the use of genomic methods and, no less valuable and important, unexpected problems that have been made visible by the same means. Genomic technologies have certainly been helpful in clarifying phylogenetic relations between "higher organisms," but when we turn our attention to microbes we find that they have been responsible for perhaps the most profound revision in biological taxonomy for over a century. This is the replacement of the traditional (and still widely disseminated) highest fivefold layer of classification (animals, plants, fungi, protists, and bacteria), with a threefold classification into Bacteria, Archaea, and Eukarya, where the first four groups in the traditional taxonomy are now all classified together as Eukarya, and the fifth is divided into two highest-level groups, Archaea and Bacteria. The main instigator of this radical taxonomic innovation was Carl Woese, and its ubiquitous acceptance testifies to the value of the molecular methods that he himself had pioneered in classifying bacteria using the 16S rRNA sequence mentioned in note 6 above (Woese and Fox, 1977).

There are, nonetheless, recognized problems surrounding the use of genomic-molecular methods in the taxonomy of microbes, including those using 16S rRNA sequences. The most serious of these arises from the prevalence of xenologues in the genomes of microorganisms, or, as it is more often described, from the widespread phenomenon of lateral or horizontal gene transfer (HGT).[10] Since microorganisms just *are* cells, a phylogeny of microorganisms will ideally provide a tree that represents the lineages of cells. But what HGT implies is that the lineages of genes, in fact of any genomic elements that can be so transferred, cannot be assumed to coincide with the lineages of cells. The problem is fundamental because where there is HGT, trees of cells may no longer be inferred

10 In view of the problems discussed earlier with the concept of a gene, it would be better throughout the following discussion to speak of horizontal DNA transfer. However, "horizontal (or lateral) gene transfer" has become the accepted term here.

from the trees of genes or genomic elements that are the most that DNA comparisons can produce.[11]

The difficulties for classification presented by HGT can be well illustrated by a now classic example, the heat-loving bacterium *Thermotoga maritima*. This organism, though generally seen as a bacterium, appears to have derived no less than a quarter of its genome from archaea. On a tree of microbial life it will appear in drastically different places according to whether it is classified in terms of 16S rRNA sequences, or the "concordant genes" characteristic of the place in the tree that fits best with its general morphology, or, finally, in terms of "phylogenetically discordant genes." These last DNA sequences displace *T. maritima* from the bacteria altogether and place it among the archaea (Gogarten and Townsend, 2005). Evidently HGT threatens to undermine entirely the enterprise of identifying separate microbial kinds and setting them out as a tree and suggests that we should see relations between microbes as reticulated, or netlike, rather than treelike (Doolittle, 2005).

The problems posed by microbes for tree-based classification are profoundly disturbing to many biologists because of the *theoretical* importance of trees. The lines of descent on traditional tree diagrams (Fig. 2) are assumed to represent evolutionary processes. The diagrams have been expressions of the dominant theoretical perspective of modern biology, the theory of evolution, and to discard trees is to make a major modification not merely to a conventional scheme of representation but to our accepted way of understanding how kinds of organisms have become what they are.[12] The classifications we have been discussing as lines of descent,

11 HGT occurs via three recognized mechanisms. The first is *transduction*, the transfer of genomic material mediated by viruses, or phages. Second, many microbes engage in a quasi sexual activity, called *conjugation*. This involves the production of a special tube-like excrescence, which is inserted into another microbe, and down which material is transferred. Third, many microbes have the capacity to capture free-floating or "naked" DNA from the environment, and incorporate it into their genomes, a process referred to as *transformation*.

12 Many microbiologists still insist on the theoretical importance of the existence of a phylogenetic tree underlying microbial classification, but in practice the question of what constitutes a microbial species is almost invariably treated in a wholly pragmatic way. The phylogenetic tree is now defended by reference to a "core" genome said to be resistant to HGT, though this appears to amount to only a few percent of the genome, and its existence is controversial (Saunders et al., 2005). Pragmatic definitions of microbial species include, for example, a 70% rate of DNA-DNA reassociation in hybridization tests of the total genomic DNA of two organisms (Roselló-Mora and Amann, 2001) or 97% identity of

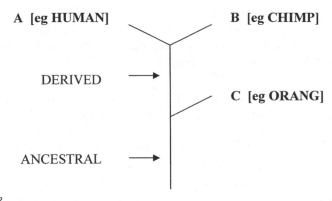

FIGURE 2

in other words, are everywhere understood as evolutionary connections; we now need directly to address the theory of evolution and the impact that developments in genomics are having on this theory.

Evolution

For many of us today there is no fundamental difficulty in understanding how the present range and variety of living things came into being. Indeed, we see the explanation as following naturally from uncontroversial biological facts. We know that all living things, including human beings, typically produce progeny more numerous than themselves. We also know that progeny vary and that this variation may be passed on in turn to their progeny, although the source of this heritable variation may elude us. Given that living things tend to multiply, some progeny must fail to survive and reproduce if the population is not to increase beyond the numbers its environment will sustain. Given that there are variations between individuals in the progeny, the chances of survival and reproduction of those individuals will surely be correlated with how they vary. Thus, through a cycle of variation, differential survival, and

---

rRNA. No deep theoretical motivation is plausible for these numbers, which seem to accord better with the view that the primary goal of biological taxonomy should be to facilitate the storing and communication of biological information rather than subservience to any broader theoretical aim, such as the articulation of phylogenetic history (Dupré, 2002, ch. 4).

reproduction we must expect every generation in a reproducing population to differ at least to some small degree from the previous one. And by imagining the cycle repeating itself over countless generations, we can provide ourselves with a wholly naturalistic explanation of how living things today, in all their complexity and diversity, may derive from the much less diverse and complex forms of life that existed in earlier times.

Everyday understandings of this sort incorporate the explanatory perspective now referred to, somewhat misleadingly, as "the theory of evolution." They assume, for example, that we humans are not merely descended from other species but that we evolved from other species through processes involving heritable variation and differential survival over very many generations. Although there has been a great deal of controversy over the details of these processes, their sufficiency, their rate of action, and much else besides, in general outline the theory of evolution is currently accepted very widely in the realm of everyday understanding as well as practically everywhere in the sciences. There is also, of course, longstanding and continuing debate as to whether the origins of life are not better understood from a theological perspective, but scientists themselves normally operate within a naturalistic framework when they study the world, and reserve any theological reflections for evenings and weekends. Insofar as they remain scrupulously naturalistic, it is very hard for them to conceive of an alternative general theory of how the current distribution of life forms came to be.[13]

This long-standing prominence of the theory of evolution in the biological sciences themselves is indeed a striking fact. Theodosius Dobzhansky, one of the leading evolutionists of the twentieth century, famously remarked that "nothing in biology makes sense except in the light of evolution," and while he scarcely counts as a disinterested, independent commentator, the extent to which his assertion continues to be cited and endorsed suggests that he was not far wrong.[14] But although it may perhaps be that we cannot make sense of current biological science without reference to evolution, it is also the case that we cannot make sense of evolution without reference to current biological science. In particu-

13 This is not the place to take up the debate between evolutionists and creationists. One of us has elsewhere attempted to explain in more detail what the theory of evolution is and is not committed to, and its advantages over theological alternatives (Dupré, 2003).

14 The ISI Web of Science records over two-hundred citations of his five-page 1973 article with this title.

lar, only by referring to the biologists who currently formulate and apply it can we understand what "it" currently amounts to. We need to be aware that the theory of evolution has not remained the same theory over its history. We tend to refer to "the theory" today as though we can take for granted what we are speaking of, and as if it is something that has been in our possession, basically unchanged, for well over a century. But if we try to make the theory explicit and set it out systematically, we are likely very quickly to realize that neither of these assumptions is correct.

Many theories of evolution were advanced in the course of the nineteenth century, but the lineage of the modern theory is generally traced back to Charles Darwin, who strove assiduously to make his theory scientifically respectable and drew extensively on existing knowledge of large living things to give it empirical support.[15] Like the modern theory as we summarized it above, Darwin's focused upon variation and differential survival. But several other more specific elements of the theory continue to be accepted, along with some of the anthropomorphic forms of language in which it was expressed, although there is more awareness of the problems raised by this language today. References to "competition" between organisms, to "natural selection," "adaptation," and "fitness" are examples here, still widely encountered in specialized contexts as well as in popular discourse, where life is still sometimes represented as a "struggle for existence."[16]

15 Darwin had the inspired notion of deploying an impressive body of empirical work in support of his idea by describing how the domesticated animals and crops that were then (and now) used by human societies were the products of thousands of years of selective breeding. By breeding animals, or the seed of plants, with particular desired qualities, plant and animal breeders had, in a relatively short span of time, wrought remarkable changes from the wild animals and plants that were their original raw materials. Given the unimaginably longer periods of time already thought to characterize the earth's history, the changes that might be brought about by the natural analogue of this process seemed almost unlimited.

16 Another unfortunately anthropocentric term is, of course, "evolution" itself, but Darwin was not fond of the term and scarcely ever made use of it. Indeed, there are even occasional hints in his work that his main scientific interest was not in evolution at all but in the origin of species. Ironically, there is not a great deal in Darwin still remembered as a major contribution to an understanding of speciation. We tend to forget this aspect of his contribution today, along with many of his conjectures on the nature of inheritance and on how precisely selection operated.

As knowledge of the biological sciences has grown ever more wide-ranging, detailed, and sophisticated, our understanding of evolution as a process has grown and developed in parallel, and "the general theory" itself has been continually reformulated. The theory of evolution has itself evolved. That is a story we cannot begin to tell here, but it is worthwhile to note the merger between Darwin's idea of evolution by natural selection and Mendel's particulate view of inheritance after a long period when these theories were supported by distinct and occasionally hostile camps. Darwinists generally saw inheritance as a blending of the traits of two parents, but as early critics, most notably Fleeming Jenkin, pointed out, this would rapidly homogenize the variation available for selection. A rare trait, or "sport," would be diluted by half in each generation, and soon disappear altogether. Mendelian particulate inheritance came to be seen as providing a reply to this objection, since Mendelian factors were thought of as being passed on in their entirety or not at all. For this and other reasons a merger between these two biological research programs began to be brokered, and by the late 1940s people were increasingly inclined to speak of a "New Synthesis" in biology.[17]

Despite the importance of the concept of particulate inheritance, which allowed evolution to be described as the spread by natural selection of randomly generated mutations, evolutionists did not abandon their favored view of evolution as a gradual, incremental process. Rather they assumed that rare small Mendelian mutations providing incremental beneficial changes were the only material on which natural selection worked in a positive sense and that gross mutations were always deleterious irrelevancies.[18] The theory of evolution accordingly became (for some at least) a theory of natural selection acting on variations in the form of small Mendelian mutations equally well described as mutant alleles. And because evolution was understood as involving the spread of novel alleles through a population, it followed that the unit of evolution was defined by the limits to the flow of genes. The unit of evolution had of course long been identified as the species, and indeed since *The Origin of Species* itself had been thought of precisely as an account of the emergence and stabilization of those units, so this led to a dominant view of the nature of the species, championed especially by the highly

17 How far there really was a new synthesis is contested. It is part of the heroic history of evolutionary theory, but its place in actual history is less clear.

18 The thought that continuous incremental change was a politically attractive version of evolutionary change naturally comes to mind here.

influential evolutionist Ernst Mayr, according to which species were defined by their reproductive isolation.[19] But, as we shall explain, the view of evolution developed by Dobzhansky, Mayr, and others has become increasingly problematic in certain respects.

Amidst the many and various further changes to understandings of evolution that have followed the New Synthesis, it is worth identifying two general directions of development and transformation. One looks away from the specific organism, and even the entire population of organisms, to both the material environment and the other species that inhabit it, and enriches the theory with ecological and other macroscopic perspectives. Recently, for example, it has become increasingly common to observe that the "niche" to which an organism is adapted is not something that existed all along, but is constantly being created and modified by the organisms that occupy it (Odling-Smee et al., 2003). The second direction of transformation focuses even more narrowly and asks how specific parts of the organism, or processes occurring within it, or cellular features in complex multicellular organisms have evolved, thereby enriching the theory with insights at the microscopic level. This second form of development is of immediate interest, and it is worth noting how it can result in changes better regarded as transformative of the theory itself than as mere modifications of it. Recall how evolution was initially understood as change in populations of organisms of a given species, due to their variation and the differential survival and reproduction of individual organisms with different characteristics. The theory took variation as given—often it was described as random—and often spoke of natural selection acting on the variations encountered in a population of organisms. But from the start it was clear that some of the things that affected this variation could be studied and described scientifically, that these things themselves varied, and indeed that they too evolved. And of course for us genomes are crucial instances here.[20]

Today, it is largely taken for granted that evolution may occur not just at the level of the species or even the individual organism but at many different levels, and that units of many different sizes and kinds may be subject to selection (that is, survive differentially according to

19 The view of species characterized in Mayr's terms was, for Dobzhansky and his contemporaries, a central part of what was required to make sense of everything else.

20 A comprehensive discussion of the evolution of the genome, on which we draw at several points in this chapter, is Gregory (2004).

how they vary).[21] Certainly, it is assumed that parts and subsystems of organisms evolve, as do single cells, even though their usual mode of reproduction apparently precludes variation, as also do gametes, genes, and, at least on some accounts, even molecules. Indeed, in the overall story of evolution molecules have come to feature as honorary organisms. According to theorists of the early origins of life, molecules evolved in the primeval soup. In particular, they evolved methods of reproducing themselves and went on to grow in size and complexity as molecular systems, until finally we were inclined to call them living things and to track their further evolution under that description. And here at last we arrive at the link between evolution and genomics that is our present concern; for genomes are molecules, and probably the distant descendants of the RNA molecules central to the self-replicating molecular systems that evolved in the primeval soup.

Let us pause here to take stock. Our standard view of evolution is of a process occurring in populations of organisms that reproduce sexually and develop and differentiate from single cells: paradigm instances might be large mammals, even humans or primates, organisms that already have a long evolutionary history. The evolution of these highly developed organisms is understood naturalistically as the result of a long process during which ancestral forms are gradually and cumulatively modified by the natural selection of small differences. We should also note that this standard view often has a teleological character, at odds with the naturalism formally asserted as the key postulate of the evolutionary account, which can easily create the misconception that "higher" animals and humans are the outcome of an evolutionary progression.

If we set aside this standard account and try to describe evolutionary change over its entire duration, a very different vision emerges. We need to start as close as we can to the beginning, with the oldest things involved, which are also the smallest things. From there we can trace a path through time, which also tends to be a path to ever larger things within and around which the descendents of the older and smaller things

21 This statement will not surprise philosophers of biology, for whom the topic of units of selection has been one of the most extensively discussed in the field. Despite some residual dissent, the pluralism we assert here has surely emerged as the dominant view. It may, however, surprise general readers, especially those familiar with the work of Richard Dawkins and his influential advocacy of the idea that only genes are selected. Here, as in much else, we disagree with Dawkins. As genomics research proceeds it is creating more and more problems for Dawkins' account of evolution.

encountered earlier in evolutionary time continue themselves to evolve. The move through evolutionary time is not uniformly and precisely a move from smaller to larger and more elaborate creatures, but it is close to being so. And thus we can set the entities that have evolved, and continue to do so, along a line from molecules, to simple prokaryotic cells, eukaryotic cells, and protists, multicellular organisms produced by differentiation and development, to familiar animal kinds like ourselves and other primates.[22] At every stage along the line, important and interesting processes of evolutionary change are identifiable, significantly different from those proposed by the standard account. And because the first objects to evolve continue to do so today, not just independently but as parts of larger organisms, reflection on the most ancient evolutionary history is also reflection on microevolutionary processes occurring in the most recently evolved kinds of creature.

In the beginning, in the chemical soup widely supposed to precede any enclosed and bounded objects we would regard as living things, the first entities to evolve must presumably have been molecules. Self-reproducing molecular systems came into being, centered, as we now believe, not upon DNA but upon RNA. This was the RNA world, now the subject of much animated speculation. This ancient history may well long remain a matter of speculation, but molecular evolution as it then existed is presumed to have continued to occur thenceforth and to be continuing today, so that comparative genomics can now treat the DNA and RNA sequences of all extant creatures as a record of evolutionary change at the molecular level. And indeed it was not long after the discovery of the structure of DNA and the advent of molecular genetics research that findings emerged about how evolution occurred at the molecular level that did not fit particularly comfortably with the tenets of the standard account.

Combining these findings with arguments drawing on population genetics and with newly emerging knowledge of redundancies in the genetic code, Mootoo Kimura (1983) proposed a "neutral theory" of molecular evolution in sexually reproducing creatures. What the evidence suggested, according to Kimura, was that evolutionary change at the molecular level involved very little in the way of natural selection and mainly consisted in functionally neutral or even mildly deleterious mutations

22 Missing from this list are viruses, either the smallest living things or, perhaps, the largest nonliving things in the story. The place of viruses in this history is controversial. This, and their evolutionary significance more generally, will be briefly discussed below.

becoming fixed in populations through chance.[23] The precise way in which this fixation was said to occur is intriguing. The production of gametes and their subsequent recombination and development into new phenotypes could be understood as a random sampling of the original genetic material in a given population, and repeated sampling of this sort would eventually lead, mathematical analysis indicated, to mutations of no functional significance either being lost from the population over several generations or increasing to a level that effectively fixed them as enduring constituents of its genetic material.[24] Fixation could be explained, that is, as the consequence of random genetic drift without natural selection being involved at all, and Kimura's (1983) analysis of what was known empirically of mutations at the molecular level suggested that drift and chance were indeed overwhelmingly the major explanatory factors and that selection was rare in comparison.

This early intrusion of molecular genetics into the debates about evolution was in many respects a modest one. It remained naturalistic and respectful of empirical knowledge. It looked to the traditional approach of population genetics for many of its arguments. It did not deny that some natural selection in the neo-Darwinian sense could and indeed did occur. However, the neutral theory did challenge the standard view that the selection of variations could be assumed to be the overwhelmingly dominant source of evolutionary change. Perhaps even more importantly, it put a question mark against the gradualism of the standard account. The neutral theory was most plausible when applied to single-nucleotide substitutions, and these were precisely the single-point mutations that best fitted the needs of evolutionary gradualism and gave it ammunition against, for example, punctuated equilibrium theories of evolution. And finally, the neutral theory insulted with indifference

23 This, of course, is the basis of arguments already discussed surrounding the possibility and reliability of molecular clocks. There are grounds for believing that point mutations involving single nucleotide substitutions occur at a more or less constant rate in suitably chosen parts of particular genomes, so their accumulation may be the basis of a molecular clock for evolutionary times if selection is not operative.

24 Intriguingly, the mathematical models relevant here are more or less the same as those used to describe technological change when alternative devices or gadgets compete in a marketplace (Arthur, 1984). The result is that one competitor may fix and the other disappear entirely because of "advantages of adoption" analogous to advantages of chance survival. Thus a "neutral theory" of the evolution of genomic technology may perhaps run alongside a "neutral theory" of the evolution of genomes.

the widely accepted teleological frame in which standard evolutionary theory was expressed, effectively exiling from center stage the notions of "fitness," "adaptation," "competition," and so forth.

It may be that these were greater sins than they first appear. Certainly, in the ensuing controversy, "selectionist" critics of the neutral theory did not simply present and interpret empirical findings to call it into question. Some of them insisted that any changes or mutations that subsequently became fixed had to have implications for fitness and had necessarily to result in some level of selective advantage or disadvantage. Selectionism, that is, was defended not simply in terms of evidence but almost as a doctrine, and the neutral theory was not allowed the standing of a competing scientific theory. Either it was treated as a priori incredible, or in a particularly interesting variant strategy, it was dismissed as a theory not of evolution itself but of mere change—an account, possibly correct, of the "noise" that accompanied "genuine" evolution (Kimura, 1983, p. 50). There is a nice reminder here of the vague and unsatisfactory nature of the notion of "evolution" itself, which makes it liable to strong semantic fluctuation (and contestation) as it is applied to processes differing significantly in their characteristics from those paradigmatic for the standard account.

Why were the changes described by Kimura not regarded as evolutionary changes? Was it perhaps because they lacked a "progressive" quality, because they did not identify the essentials of a process of change leading on to ever "higher" forms of life, culminating in us? Might some of its critics have reckoned that there was no "genuine" evolution going on here, just lumps of stuff, banging around and recombining into different forms any one of which was as good as any other? All this could well have been the case. It seems clear that for some of its proponents, including some scientists, the "incidental" teleology of evolutionary language is actually far more than incidental, that it is in truth what evolution is "really about." And the naturalistic materialist orientation encouraged by a focus on molecules clashes more strongly than first appears with the standard account to the extent that this is the case.

Let us set aside the clash of teleology and naturalism apparent here, however, and return to our more general point: the unit of evolution focused upon is liable to affect the idea of what evolution is, whether superficially or profoundly, and in far-reaching and controversial ways. This seems to have been the case when the molecule was addressed as the relevant unit and, as we shall now go on to show, a comparably profound effect results when microbes, the next step along from molecules on the line of evolutionary change, are made the main focus of attention.

A vast range of microbial life occupies a place (and great tracts of evolutionary time) between the bare molecular systems that we imagine preceding or constituting the very early stages of life and the organisms that provide the basis for the standard view of evolution. And microbes differ from these more recently evolved organisms in respects that are fundamental to the standard view. Microbes do not have sex,[25] and as the phenomenon of lateral gene transfer makes clear, their evolution is not constrained within the boundaries of the reproductively isolated species. These features of microbial life make the notion of species that appears in the standard story inapplicable. And in the absence of sex, the standard Mendelian models of inherited variation have no direct application either. As we have noted, most of the history of life is the history of microbes, and for this part of the history we need a model of evolution in which microbes are the paradigmatic organisms.

Between the world of molecules, in which self-reproducing systems somehow come into existence and perhaps evolve, and the world of microbes, a crucial event is generally reckoned to have occurred, a "Darwinian transition" involving the confinement of chemical systems and the emergence of life as bounded living objects or cells. A precondition of the Darwinian transition was the development of the cell membrane, the structure that allowed the metabolic processes of organisms to occur largely insulated from interactions with whatever lay outside the cell membrane. This facilitated the emergence of cellular lineages sufficiently isolated from the general genetic free-for-all for the discrete evolutionary processes to occur that are often taken to be a prerequisite for the evolution of the complex and diverse organisms that have followed over the last three billion years.

The organisms that launched the Darwinian transition are generally thought to be the ancestors of contemporary bacteria. Genome sequencing of a substantial number of kinds of bacteria has supported the hypothesis that basic processes of DNA replication are common to all extant bacteria, and were very probably established contemporaneously with this emergence of cellular life. We can presume that like most extant bacteria, these first cells were of relatively simple form, consisting of a single boundary membrane enclosing a volume of cytoplasm, and enclosed within a semirigid cell wall. Cells of this sort are generally referred to as prokaryotes and contrasted with the much more complex cell types found in eukaryotes—plants, animals, fungi, and a variety

---

25 One method of HGT, conjugation, is often described as mating, however. This is explained below.

of single-celled organisms referred to generically as protists. Prokaryote cells lack the complex internal structure of membrane-based compartmentalization, including a distinct nucleus, found in eukaryote cells.[26] They also lack the chromosome structure found in eukaryotes, and typically have most of their DNA arranged in a single circular structure. It is commonly supposed that the earliest eukaryotes evolved from an ancestral prokaryote, or perhaps the merger of two ancestral prokaryotes, possibly a bacterium and an archaeon. At any rate, the process by which this happened remains highly controversial.

As we have said, for most of the period that separates us from the Darwinian transition, prokaryotes were the only living things, so that the story of how life has evolved ought to begin with the evolution of prokaryotes. But in truth this topic is striking for its absence from many of the most influential discussions of evolution. Most evolutionary theorists, though certainly not all, have been interested primarily in the period since the development of multicellular organisms[27] and have given scant attention to the microbes that, as well as being the only life forms for most of evolutionary history, remain the predominant form of life by far even today. Despite their microscopic individual dimensions, microbes are estimated to comprise over half the total biomass of life on Earth, which makes their neglect in the context of evolutionary biology a truly remarkable feature of the history of science.[28]

Two further points will serve to emphasize just how myopic this neglect is. First, as we have already discussed, the neglect of microbes has led the standard account to give undue emphasis to the evolutionary role of sexual reproduction in genetically isolated lineages: microbes

26 As with so much in biology, this contrast admits of exceptions. The Planctomycetes are a group of bacteria with a membrane-bound nucleus and other cellular structures.

27 There is no widely accepted term for multicelled organisms. All multicelled organisms are eukaryotes (organisms whose cells include a distinct nucleus), but the eukaryotes include a number of single-celled organisms, protists. For most purposes of the present discussion, protists should be treated with other microbes, the bacteria and archaea. One of us has earlier advocated the use of the word "macrobe" to refer to what are traditionally thought of as multicellular organisms (O'Malley and Dupré, 2007), and we shall occasionally use this term in this text. Later we shall point to some difficulties with the whole concept of multicellularity.

28 The failure of philosophers of biology to pay due attention to microbes and the problems this has caused are discussed in more detail in O'Malley and Dupré (2007).

are neither sexual nor in general are their lineages genetically isolated. Second, criticisms of the standard account are themselves liable to be weakened by this very same neglect. Thus, the standard account is now increasingly criticized for its failure to address the evolution of developmental mechanisms.[29] One such critic, Jason Scott Robert, has written, "Development is what distinguishes biological systems from other sorts of systems, and it is the material source of evolutionary change" (2004, p. 34). It is clear that something has gone amiss here. For development, as conceived in this context, is the process of cell differentiation and specialization by which the complex ordered systems that are mature multicellular organisms develop from a single fertilized egg; far from being distinctive of biological systems, it is actually uncommon evolutionarily speaking. Multicellular organisms form quite a small part of even the fairly recent history of life and not the totality of biological systems. Robert's just quoted remark is wrong.[30]

One way of looking at the distinction between evolution before and after the multicellular transition is to see the earlier phase as primarily the evolution of chemical machinery and the later stage as the evolution of development. Though this is of course an oversimplification, it highlights two important points. The first, which we have just mentioned, is that development, as generally understood, begins only when multicellular organisms appear.[31] The second is that most of the chemical machinery of life existed before the transition. Mechanisms of DNA transcription and translation, and the "code" relating nucleotide triplets to amino acids, for instance, exist in both eukaryotes and prokaryotes in essentially the same form, though with some more or less subtle differences. As will be further emphasized later in this chapter, when we discuss the mutualistic relations between microbial communities and macrobes, microbes remain the specialists in chemical machinery more than their more structurally complex descendants. We do not mean to suggest that evolution of chemical machinery stopped two billion years

29 This failure can in part be traced to a tendency to think of genes as somehow encoding phenotypic traits, that is as Genes-P in the sense defined by Moss (2003). This tendency has made it easy, perhaps inadvertently, to "black box" the developmental program.

30 Despite this reservation we have considerable sympathy with the so-called evo-devo criticisms of mainstream evolutionary theory.

31 Microbiologists do speak of development with relation to microorganisms, referring to sequences of stages through which they pass, but this is, of course, a somewhat different meaning of the term.

ago, but only that recent developments may be regarded as relatively minor compared to the earlier achievements of our microbial ancestors (and cousins). This may be related to the fact that, given what we have said about the failure of microbes to exist in neatly isolated cell lineages, for most organisms the Darwinian transition never really happened.

Let us now focus once more on evolution at the molecular level, the level involving genomes. Genomes mostly evolved in the context of a role within microbes, and we need accordingly to address the differences between their evolution in that context and their evolution within the large sexually reproducing organisms of the standard account. We have already mentioned the distinction among elements of genomes between orthologues, paralogues, and xenologues. Microbial genomes typically include elements of all these kinds, and this implies a more complex picture of the evolutionary origins of any particular genome than the usual picture based upon the standard macrobe-centered view of evolution. Particularly important to this more complex picture is the role of the xenologues characteristic of, although certainly not unique to, prokaryotes.

Xenologues, by definition, have arisen from the transfer of genetic material across lineages or, where they exist, species boundaries. This, as noted above, is commonly referred to as horizontal, or lateral, gene transfer. HGT has generally been considered an uncommon occurrence between eukaryotic species, and not something that significantly increases the genetic resources available to them.[32] But this assumption is now becoming more controversial, and substantial proportions of eukaryotic genomes are currently thought to have originated by HGT. There are a number of ways in which HGT may be effected, but viral insertion is now believed to have been the most important source of xenologues in eukaryotic genomes, as it may perhaps also have been in prokaryotes.

Viruses are the simplest biological organisms, containing a relatively small complement of either RNA or DNA and an external protein structure. It is sometimes denied that they are biological organisms at all, and it is true that unlike (other) living things they can exist in static crystalline states and that they need to hijack the reproductive machinery of other organisms in order to reproduce themselves. Whether viruses are

---

32 This is despite hybridization between related species being recognized as a common phenomenon, so common indeed that the ensuing taxonomic problems are routinely avoided by insisting that a group of putative species sufficiently similar to allow genetic transfer through hybridization should be treated as one larger polytypic species.

or are not living things seems to us a matter of little moment. More significant is that our understanding of their role in biological processes
has been undergoing serious revision in recent years, and that it is now
conjectured that they may have an absolutely central role (Villarreal,
2004). As with cellular microbes,[33] our insight into the nature of viruses
has been transformed by the availability of genomic technologies. Virus
genomes are small and easy to sequence, and at the time of this writing
some 1,500 have been sequenced (see http://www.ebi.ac.uk/genomes/
virus.html). One question that has been reopened by this new data is
that of where viruses come from. It used to be generally believed that
they were degenerate remains of more complex organisms. It is a familiar observation that obligate parasites tend to lose functions as these are
carried out on their behalf by hosts, and viruses may be hypothesized to
be parasites that have handed over their reproductive functions in this
way. However, a quite different opinion is gaining ground that viruses
may actually be very ancient life forms, even tracing their ancestries to
the RNA world we mentioned earlier. Allied to this suggestion is the intriguing speculation that viruses may have played a central role in the
evolutionary process.

All viruses exploit the replicatory machinery of their hosts, and in
some cases they do this by inserting their own DNA (or transcribing
their RNA) into the host genome. There is an obvious potential for this
process to provide lasting xenologous additions to the host genome.
Of course, it may not always do so. If the host organism is a complex
eukaryote, the added genetic material must find its way into the germ
cells, which may be an obstacle that makes such additions rare in these
organisms. This possible obstacle does not exist in single-celled organisms, however, whether they are prokaryotes or eukaryotes. All cellular
microbes are to some extent parasitized by viruses, or phages, as their
viral parasites are generally known.[34] But prokaryotes generally have extremely dynamic genomes, and they are extremely good at excising alien
DNA as well as at letting it in. So the extent to which accumulation of
inserted viral DNA occurs as an important factor in microbial evolution
cannot be decided on general principles and will become clear only as
relevant evidence accumulates.

---

33 Viruses are sometimes thought of as a kind of microbe, hence the reference to other microbes as "cellular microbes."

34 It has been estimated, for instance, that 10–30% of the oceanic population of bacteria are destroyed by phages daily! And by no means all phages are
lethal.

For the purposes of this chapter we point to this question only as a major issue with almost unlimited potential to transform our view of the evolutionary process. Thus, Patrick Forterre (2005, 2006) argues persuasively that the transition from an RNA to a DNA cellular economy may have involved distinct DNA replication machinery being added to RNA cells at the origins of each of the three domains of life (bacteria, archaea, and eukarya), and that in each case this was the result of transfers from DNA viruses to RNA cells. And more generally Villarreal (2004) suggests that viruses may be the major source of evolutionary novelty. He points out their astronomical numbers—oceanic viral populations, for instance, are estimated as being on the order of 10^{31}, a number unimaginable to most of us and one that, combined with extremely rapid generation times, points to the existence of a uniquely large evolutionary potential.[35] Villarreal also notes that sequenced viruses often reveal a very high proportion of genes not known in the cellular world, suggesting that there is a large reservoir of biochemical variation here. Comparing the familiar population sizes and generation times of "higher" organisms with populations of viruses in the trillions or higher and generation times of a few hours, it is easy to see that the exploration of chemical possibility will occur overwhelmingly more rapidly in populations of viruses. And the combination of this with a known mechanism for insertion into recipient genomes invites a number of tantalizing speculations about the evolutionary role of viruses, several of which may soon become more than that as the quantity of comparative genomic data increases and our tools for analyzing it improve.

Metagenomics

Whether as a result of viral activity or otherwise, the importance of HGT in single-celled organisms and especially prokaryotes is now hardly open to debate. Indeed, recognition of the prevalence of HGT within microbial populations has led to one of the most remarkable conceptual developments to have emerged from recent genomics. The scientific project known as metagenomics first came to prominence with the yachting expedition of Craig Venter, well-known as one of the main contributors to the Human Genome Project (HGP), following the route of *The Beagle* to collect microbial specimens from around the oceans of the world. The first result of this project was a collection of microbial DNA from an

---

35 Human viruses such as HIV evolve rapidly even in the context of a single host, which is a serious obstacle to developing effective therapies or vaccines.

oceanic microbial community in the low-nutrient Sargasso Sea (Venter et al., 2004). The inventory of DNA that this generated was at the time the largest genomic dataset that had been catalogued for any community. Almost 70,000 previously unknown genes were discovered, and 148 previously unfamiliar elements in the 16S rRNA sequences often used to define microbial species. Special provisions had to be made in the main genomic database (GenBank) so that the data did not overwhelm all the more standard genomic data contained in the databank and skew subsequent comparative analyses (Galperin, 2004).

Although this result was fascinating enough merely in terms of the biological diversity it disclosed, the term "metagenomics" can also refer to an even more intriguing theoretical idea. We have noted the ease with which microbes exchange DNA fragments with one another or assimilate DNA from the environment, for example from the remains of organisms that have been destroyed by viruses. Microbes are most often found in complex and highly interdependent communities, and the rate of DNA exchange is at its highest within the members of such communities. This has led to the idea that rather than thinking of their genomes as the exclusive property of individual organisms, we should think of a *metagenome* encompassing all the genomic resources available to a microbial community. Genetic exchange within such communities, whether or not it is specifically provoked by environmental contingencies, is clearly often functional. One example of this is the well-known phenomenon of antibiotic resistance. The resources to neutralize antibiotics (and presumably other naturally occurring toxins) are often found in what are known as gene cassettes, segments of DNA flanked by units that are specifically adapted for insertion into specialized sites in microbial genomes referred to as integrons. This mechanism gives microbial communities effective mechanisms for distributing such highly adaptive resources among their members.

The investigation of metagenomes has become a fairly well-established aspect of microbiological research, and there is now nothing untoward about making reference to these scientific objects. What is more controversial is the thought that the community that shares this set of genomic resources should itself be seen as an individual organism, a metaorganism with a metagenome, as it were.[36] Microbial communities do indeed display many of the characteristic features of single complex organisms. Consider, for example, biofilms, the complex communities in which the

36 The argument for this proposal is developed in more detail in Dupré and O'Malley, 2007.

large majority of microbes in the wild are to be found. Biofilms can be discovered on almost any moist surface, from the slippery coverings of underwater rocks to the surfaces of teeth (dental plaque) or of catheters or drinking-fountain valves. They typically consist of a variety of distinct kinds of microbes—dental plaque, for instance, contains about 500 kinds—with distinct functions (for example, there are bacteria that specialize in adhering to surfaces), and with characteristic developmental trajectories of sequential colonization. It is becoming clear that these microbes have a wide array of means for communicating with one another and thus coordinating community behavior. The best studied example of communicative interaction between microbes is so-called quorum sensing, the ability of microbes to change their behavior in response to the size of the communities of which they are part (Bassler, 2002). However, it is becoming increasingly clear that this is the tip of the iceberg, and that microbes use a wide variety of chemical signals that moderate one another's behavior. These signals provide the essential underpinning for the most general characteristic that identifies something as a whole rather than a casual assembly of parts, the subordination of the behavior of parts to the well-being of the whole (a characteristic that has led some to refer to social insect communities, for example, as superorganisms), and this criterion appears to be satisfied by many microbial communities.

Perhaps the strongest argument for considering microbial communities as individuals, however, comes from redirecting our attention to macrobes. It has become common to think of an individual organism as a set of cells, each possessing a specific, unique genome. Thus, for example, the largest organism in the world is said to be a North American honey fungus, up to 8,500 years old and covering nearly 10 square kilometers of forest floor. It may not look like a single organism, but what appear as individual mushrooms are in fact connected together by a vast subterranean tangle of mycelium, all of which together constitute a single monogenomic cell lineage. More generally, and putting on one side some important qualifications ranging from somatic mutation to epigenetic differences between cells in different parts of an organism, it is true that complex eukaryotic organisms develop through a process of replication and differentiation of genomically similar cells. However, all, or certainly almost all, such organisms depend for their proper development and thriving on interactions with a variety of other, usually microbial, organisms, and the custom of considering the genomically similar cells to constitute a discrete individual but the microbial cells to be only contingently associated fellow travellers needs to be spelled out and defended. The particular evolutionary perspective in which it is usually grounded, according to which evolution is assumed to occur in isolated lineages, might be thought unduly limited in view of the phenomena of DNA exchange between lineages and of intimate cooperation between members of distinct lineages that we have just described.

A human body, for instance, carries around with it a vast number of mutualistic—that is, mutually beneficial—microbes constituting, in fact, about 90% of the cells in the human body (Savage, 1977) and, for that matter, containing some 99% of the genes in a human body. The majority of these are found in the gut, but they are also on the skin, in the upper respiratory tract, and in the various other bodily cavities. Although the extent of the mutualism involved is not yet well understood, it is increasingly clear that these microbes play central roles in digestion, immune response, and even development. Animal models in which mutualistic microbes have been eliminated do poorly and develop abnormally, and in one model, the zebrafish, the presence of gut microbes has been found to affect the expression of several hundred genes (Rawls et al., 2004). A similar mutualism is found in plants. In fact, at the interface between a plant's roots and the surrounding soil are found some of the most complex symbiotic systems on Earth, mediating the uptake of nutrients by the plant from the soil. These include large consortia of microbes within and outside the roots, and fungi within the roots that extend mycelia into the outside soil. Some of these associations are known to be essential for the plant's survival, including the well-known association between leguminous plants, such as peas and beans, and the nitrogen-fixing bacteria that inhabit their roots, but it is doubtful that any plant could survive outside a carefully regulated context such as a hydroponic culture without some association with commensal bacteria.

Reflection on findings of this sort suggests that we could if we wished develop a much more inclusive concept of the multicellular organism than has been traditional. The kind of organization and coordination of cells found in organisms now treated as multicellular, for example, birds, trees, fungi, and sponges, are already extremely diverse, and we have long been familiar with such things as lichens, symbiotic associations between fungi and a photosynthetic single-celled partner, either algae or cyanobacteria. It might be good to commemorate in our very concepts the recognition that cells form a vast variety of associations both with genomically similar and genomically diverse partners, and there is no evident reason why we should not do so by identifying such associations as individual organisms. Reality is not going to mind one way or the other. Pragmatic considerations offer little in the way of guidance.

And not even our special interest in bounded objects of particular evolutionary significance counts against such a move. Certainly, there are grounds for holding that microbial communities are bounded objects with just such significance. And it is doubtful whether any important purpose, whether pragmatic or theoretical, is served by refusing to identify the heterogeneous associations of cells that constitute the bodies of complex creatures as single organisms.

Of course, none of this amounts to a decisive argument against existing systems of classification and existing taxonomic practices. In the last analysis, how we classify the biological world is something to be agreed upon, and what currently appertains is what we have so far agreed upon. There are several viable ways of breaking up life as we know it into so many individual organisms, just as there are different ways of sorting those individual organisms into kinds. What our discussion of the impact of genomics actually does, therefore, is emphasize that there are alternatives to what has gone before, that the favored approaches we have inherited are historical choices reflecting specific concerns and interests, and that in a context increasingly being transformed by molecularization and new technologies, not to mention the profound secular political and cultural changes occurring in the wider society, significant modifications to the status quo are very likely to occur.

Even so, it is striking how easily molecular and genomic perspectives suggest ways of ordering the biological world profoundly opposed to those built into the most widely diffused and accepted accounts of evolution. Since the time of the "New Synthesis" these accounts have been very strongly atomistic, both in their Darwinist view of the individual organism and their Mendelian view of the unit of inheritance, whether gene or allele. And while they have tended to stress their naturalism, and often their materialism, by way of contrast with the anthropomorphism and teleology of the religiously inspired narratives that they took to be their major competition, in fact they have tended themselves to manifest both of these characteristics.[37] They have concentrated on complex creatures close to ourselves in evolutionary time and easily understood as precursors of ourselves, and have favored narratives of individual competition, fitness, adaptation, and so forth unduly dependent semantically on prior understandings of human actions and social relationships.

37 It is tempting to speculate that competition for audiences that demanded narratives framed in these terms, or that found it difficult to think without recourse to them, is what has led to their preservation in modern ostensibly naturalistic biological accounts of evolution.

In contrast, a genomic/molecular orientation directs attention easily and naturally away from ourselves, even to the other end of evolutionary time, and firmly discourages any residual tendencies to understand evolutionary change backwards as something directed towards where we currently stand. It is able to conceptualize processes of evolutionary change in terms of exemplars of chemical change rather than social change. And it avoids the atomism of previous accounts, both at the level of the individual organism, where it has long faced difficulties, and at the level of the unit of inheritance, where its most serious problems have been widely recognized only quite recently. We should say here to anyone who sees little difference between atomistic and molecular approaches that in truth there is all the difference in the world.[38] It may be that microbes are both the most important organisms as far as evolutionary change is concerned and the first objects that can be described as living organisms. But molecules as much as microbes require the structural, systemic, and holistic modes of perception and cognition sometimes said to be characteristic of the biological sciences, and are no more intelligible in atomistic terms than organisms are.

No doubt many defenders of standard accounts of biological evolution, particularly when they address nonspecialist audiences, will continue to see religious narratives, such as those currently promoting the idea of intelligent design, as the main challenge to their position. And they could well be right to do so, particularly perhaps if they live and work in the USA. But they may also have to face an increasing challenge from within the sciences themselves, from critics articulating accounts at least as materialistic and as naturalistically inclined as their own, if not more so. Indeed, it could be that the main changes in our understanding of evolution over the next one or two decades will involve the dissolution of an older less-than-satisfactory form of materialism in the face of a burgeoning molecular materialism and a shift of interest away from those organisms that most resemble ourselves toward those that are most reproductively successful.

38 The accusation of "reductionism" tends to be evoked in both cases, but whereas an atomistic treatment of objects as ensembles of immaculate essences, both indivisible and unchangeable, may perhaps be validly criticized in this way, such criticism would simply represent a failure to understand what molecular/genomic orientations involve.

5

Genomics and Problems of Explanation

Astrological genetics and explanatory genetics

Even in our ever more utilitarian times, genomics seeks as much for explanation and understanding as for predictive success, and the same has always been true of genetics. Classical geneticists sought both to describe distributions of phenotypical traits and to explain them in terms of genetic causal antecedents, and human behavioral dispositions and psychological propensities figured prominently among the traits that interested them. But this component of their inheritance has been more of a liability than an asset for the new genetics and genomics. It has cast a shadow over how they are perceived, and it has adversely affected how they are evaluated by many audiences. Furthermore, the efforts of the new sciences to move this kind of research forward have yet to produce generally accepted results of any great significance.

Classical genetics emerged as a science in contexts obsessed with descent and ancestry as badges of status and worth, and interacted with everyday accounts of these things in interesting and complex ways. Its ideas and conjectures were taken up in these contexts by audiences keen to draw specific political and moral implications from them and happy to simplify and reconstruct them to suit this purpose. Historians have documented the unfortunate consequences of this in several contexts, describing how genetic theories widespread early in the last century were deployed to explain putative psychological differences between sexes and races, and how hereditarian explanations of the social distribution of intellectual powers and propensities provoked the heredity/environment controversies that proliferated after the Second World War and continue to this day. Accordingly, in this chapter on problems of explanation, we give more attention than elsewhere to genetics as it has been understood by nonspecialist audiences and to some of the widespread myths and parodies that have evolved of what genetic explanation involves. And we give particular prominence to human psychological and behavioral traits, since it is over the explanation of these that the shadow of the past looms largest.

Needless to say, the shadow extends well beyond the bounds of specialized science. It currently extends to media that now more than ever are explaining our behavioral dispositions in terms of a facile genetic determinism. This determinism can take extreme forms: our entire moral nature, and the fate it brings upon us, may be attributed entirely to our inheritance, and our futures read off from our genes much as they are read off from our stars by astrologers. The press in particular likes to speak of genes and heredity in this way, but its readings of our genes are not as anodyne as its readings of our stars. Astrological genetics in the press refers to "genes for" politically sensitive traits, like intelligence, criminality, homosexuality, addiction and substance abuse, and propensity to violence. Branded as the properly scientific way of explaining such traits, these accounts encourage misguided ways of responding to them and to those groups or individuals supposedly carrying the genes for them.

Specialists must always bear some responsibility for how their ideas are understood in the everyday world, but classical genetics cannot be held wholly responsible for the simplistic genetic determinism so often imputed to the field. Indeed, proper examination of the field and its history would serve admirably as a prophylactic against genetic determinism of that sort. The familiar stereotype of classical genetics is a myth, currently endorsed by both purveyors of astrological genetics and their critics. For the former, traditional genetics is the science that legitimates their own facile discourse, while for some of the latter it is the carrier of a discredited deterministic ideology with which an allegedly holistic, nondeterministic genomics having more liberal and egalitarian political implications can be contrasted.

Myths and stereotypes structure our present understanding of history every bit as much as they structure history itself. We are too much inclined to understand classical genetics as a precursor of astrological genetics. But myths and stereotypes are rarely purely fantastical, and there is no denying that the some of the explanations proffered by classical geneticists were consistent with this widely credited myth of the field. And even in genetics and genomics today, after a period when links to their past and to genetic determinism in particular were a source of anxiety, specialists of various kinds even including researchers themselves can be found deploying the discourse of astrological genetics, especially in order to draw the attention of outside audiences to their work and to obtain financial support for it.

Neither a misreading of their historical roots, however, nor a misunderstanding of how they currently account for behavioral traits that has been abetted and reinforced by some of their own less attractive discourse should be allowed to blind us to the genuine explanatory potential of the new sciences. Over the long term, they could transform our understanding to such a degree that just about every human trait becomes linked to genetic/genomic causal antecedents in some way or another. But even as the scope of genome-linked explanation is extended, so also, it is to be hoped, will our sense of its insufficiency increase and provide the necessary counterweight to astrological genetics and its overblown genetic determinism. Natural scientists, including geneticists and genomicists, do sometimes tend toward a deterministic conception of the natural world. But the contrast between a properly articulated scientific determinism of that sort and the determinism of astrological genetics is vast. The former engenders awareness of how very many different causes and kinds of cause have to be taken into account in explaining any particular empirical event, and of how any specific causal story must be incomplete and any attempt at prediction enveloped in uncertainty. The latter is more about generating supposedly secure predictions from the scan of a few beads.

Sadly, however, merely to expose the defects of astrological genetics is unlikely to bring about its demise. As well as the strange attractions of the fatalistic explanatory scheme it shares with phrenology and palmistry, more reputable considerations broaden its appeal. Sometimes it does provide quite acceptable explanations: in many human single-gene disorders, as we have noted in earlier chapters, a simple Mendelian account of how the allele for the relevant disease was acquired may serve well enough for explanatory purposes. And the simplicity of explanations of this sort is indeed another of their assets, and stands in striking contrast to the complications we shall face as we attempt to move beyond them. Unfortunately, although the problems associated with efforts to produce genetic explanations of human behavior are too important to pass over, simplicity does seem to be on the side of the devil in this context.

Heritability

The level of criticism of genetic accounts of human characteristics is lower than it once was, but considerable hostility to this kind of explanation remains, especially where human psychological dispositions and abilities are concerned. Memories of the old eugenic programs and their invocations of genetic themes have persisted over the decades. These programs were used to justify discrimination against particular groups in terms of inherited human differences, citing inferiority in nature to justify inferiority in treatment, and mental powers and capacities were usually the most salient components of human nature here. The conviction that little could be done to change inherited characteristics reinforced the inference from different nature to different treatment. But it is easy to see that all the connections cited here could have been made differently, since many of them are made very differently today. For example, the extent to which we now permit the biological differences between men and women to justify differences in how we treat them is far less than once it was, at least in most of the developed world.

Eugenics as it existed in the first half of the last century is no longer advocated outside of the lunatic fringe. Nevertheless, research on the genetic basis of human behavior is still reckoned to have moral and political implications, and is still often structured by powerful political pressures and engulfed in controversy as a result. Researchers wishing to emphasize the genetic basis of human characteristics avoid dwelling on the insufficiency of favored explanations, or their conjectural character. Critics tend to highlight what researchers downplay and to point to the many possible alternative causes of variation in behavior other than genomic/genetic ones. The term now used to refer to this array of alternative causes, from the intercellular to the geopolitical, is "the environment," and the resulting exchanges constitute the heredity/environment controversies that continue to be a feature of our intellectual and political life.

It is only by acknowledging their salience to political questions, understood in the most general sense as questions about how we ought to treat each other in the living of our lives, that we can begin to understand why heredity/environment controversies have continually punctuated the history of genetics. Considered in the abstract, they are strange controversies. No geneticist denies the relevance of "the environment";

no critic discounts in advance the role of heredity. Most of those involved in the debates appear to have a common framework of understandings allowing a role for both, something occasionally referred to as "the interactionist consensus." But let an empirical study appear, claiming to throw light on how far, in some given environment, variation in intelligence, or some other valued or disvalued trait, is down to heredity and how far to environment, and a furious altercation can practically be guaranteed. The resulting literature reflects neither the technical accessibility nor the theoretical interest of the traits studied, and is concentrated instead on characteristics that have high social and political salience, and upon their social distribution, most notoriously between the sexes or between different "races" or "ethnic groups."

It is worth noting at the outset that the relevance of much of this work to public policy is far from obvious. Take the great quantity of work published in educational policy journals in the USA on the high heritability of, and alleged racial differences in, IQ. Intelligence is considered important in the US as elsewhere, and higher educational institutions try to recruit a proportion of students not unduly lacking in it. But nobody suggests today that institutions should (or could) maximize IQ in their intake by recruiting on the basis of race, and to the extent that maximizing IQ is indeed the intended goal, it would be a simple matter to give all applicants an IQ test. Given also that the measures of heritability we are about to discuss have no implications for whether the IQ of a group could be raised, or a between-group difference in IQ narrowed, by the activities of educators, it is unclear to say the least why such measures should have any policy relevance.

One genuine and valuable outcome of the heredity/environment debates is an awareness, at least among relevant experts, of the enormous technical difficulties involved in separately identifying the importance of these two factors. Another is an increased recognition of how problematic some of the terms and classifications most prominently deployed in accounts of human biological variation are. As often as not the categories of person involved are closer to being bureaucratic inventions than natural kinds, the products of self-reporting on forms and documents and the mechanical aggregation of ticks in boxes rather than of external observation or ascertainment. This is true of race in particular, certainly as the notion is employed in the USA (see, for example, Dupré, 2008). Nor are the trait variations referred to in all cases readily observable empirical characteristics. IQ test scores, for example, have an unduly artifactual quality, linked to their role in legitimating various kinds of administrative ordering and sorting, and the question of just what it is that is measured by these putative tests of "intelligence" continues to be raised. And some of the categories used to describe human pathologies, especially mental illnesses like schizophrenia, are notoriously problematic as empirical descriptors and are not always applied consistently even by trained diagnosticians. A recent study of the genetics of schizophrenia in African-Americans, for example, cites among its aims that of remedying what according to clinical evidence is the very substantial overdiagnosis of the condition in this group.[1]

Difficult issues arise here, most of which we must pass over, but it is important to keep their existence in mind since they imply the need for caution in evaluating claims about human variation and especially human behavioral variation. Certainly, claims about the correlates of racial variation always need to be treated with reserve, as serious problems surrounding the notion of race and its appropriate use are going to persist and to affect even the most well-intentioned scientific and medical research. Whenever the incidence of illnesses and pathologies is studied in different populations, significant differences tend to be found, and in many such cases a probable link with genetic variation can be established. And similar links are now increasingly being found when variation in the efficacy of treatments of illnesses and pathologies is studied across populations. Moreover, these may include populations identified not by biological means but by reference to culture, language, geography, or bureaucratic records. Among them are populations conventionally described in terms of racial categories, members of which are accessed for research through their own racial self-categorizations, and accessed for any resulting treatments through administrative systems that employ those same categorizations. From the perspective of medical research, opportunities to improve health exist here: racial categories may serve as *markers* of persons and/or families on whom searches for distinctive alleles may be focused, and for whom especially effective drugs and therapies may be formulated and rapidly dispensed, even though the categories are not biological categories.

In the context of medicine and medical genetics research, interest in generally deployed racial categories is largely confined to their role as markers or pointers, sometimes as markers of populations thought

---

1 The Project Among African-Americans to Explore Risks for Schizophrenia [PAARTNERS] at www.utmem.edu/psychresearch/. There are also problems in agreeing what the condition consists of, and fascinating questions concerning why there is overdiagnosis. Clinical evidence claimed to show both a genuine association between schizophrenia and racial group *and* overdiagnosis also exists.

to be slightly richer than others in given traits or susceptibilities even if the vast majority of the individuals in the populations do not manifest them at all. And even this technical interest is increasingly recognized as a temporary one: race-linked medicine and medical research will cease to be useful, it is said, when sufficient knowledge becomes available to sustain a personalized medicine, using individual genotyping, that will meet medical imperatives better, as well as those of individualistic market societies. But even in societies where racial categories are already very widely recognized and deployed in all kinds of different ways, criticism of "benign" research of this sort and the terms in which it has been framed is to be expected and has indeed been made. To associate a group with a specific heritable illness may perhaps help toward the improvement of its health, but it may also adversely affect how the group is perceived and the respect its individual members are accorded in the course of their social life, especially perhaps where a mental illness such as schizophrenia or bipolar disorder is involved. And this is just one problem among many that ultimately derive from the fact that racial categories are generally used to refer to self-recognizing status groups in society and not groups systematically identified by reference to genetics, ancestry, or indeed biological traits of any sort.[2]

On the face of it, problems of this kind may be dealt with by clarification, by making it explicit, for example, when researchers are using racial categories inscribed in bureaucratic records and why this can be important. It is surprising that in the vast literature that reports the results of comparative medical and biological research carried out on bureaucratic kinds rather than biological kinds and on groups of humans ascertained in terms of their status rather than their biological characteristics, so little is said on these matters. It needs to be stressed, however, that clarification alone is unlikely to resolve problems that arise from researchers sharing terms and partially overlapping their concepts with those used by quite different groups and subcultures. Where such sharing exists, other users may justifiably take a critical interest in what specialists are doing with a term; indeed, with "race" there is a case for

2 These are problems that those intent upon securing equality of standing and respect for different racial groups are especially aware of. But it is an interesting sociological question why, in contrast to the case of races, efforts to secure equality of treatment and respect between the sexes have made impressive headway without "natural" differences becoming much of an issue. It is true that IQ test scores for minors in the UK were long fixed to disguise the superior intelligence of females, but how far that is relevant to the present discussion is moot.

others regarding specialists as secondary users of *their* term, who would do better to develop their own distinctive terminology. And more generally, whenever specialists develop ways of using a term that differ significantly but not completely from those in everyday life, they should ask whether they have a duty to change terminology in order to avoid any confusion.[3]

It does help, when seeking to relate human phenotypical variations to heredity and environment, if actual phenotypical traits are addressed and if populations are defined biologically. Provided that this is done, however, there need be nothing wrong in principle with projects of this sort. Indeed, there are several good technical accounts of their possibilities as well as their pitfalls. For our present purposes, however, discussion of a very simple nontechnical example will suffice. Imagine someone buying a few packs of flower seeds, and scattering them here and there in the garden in spring. Some weeks later a display of color brightens the back of the house. There will of course be discernible similarities and differences between any two individual plants in the display. But it will be striking just how alike are the plants from a single seed packet, and how different these plants are collectively from plants grown from other seeds. It is natural to assume that something contained in the seeds explains this similarity, and we now locate this something in the genomic DNA inherited by the plants. Heredity, we say, is responsible for the distinctive similarities between plants grown from the same seeds. Continuing to look around the garden, from the sunny to the shady bits, from damp areas to dry, we are also bound to be struck with how the distinctiveness of a given kind of plant persists, how the poppies remain defiantly red, for example, wherever their seed falls. This suggests that whatever goes on in the seeds is to some extent unaffected by environmental variation. On the other hand, what happens will certainly not be *entirely* unaffected by environmental variation. Typically, some seeds will have thrived while others will have failed, having landed on brick or stone; and the same seeds may have produced larger or smaller plants according to the light or humidity. Environmental variations may affect seeds of different kinds very differently, so that, for example, poppies may be uniformly red everywhere whereas the flowers of other species may have their colors fade in the sunshine, or range from blue to pink

3 Not even this, however, guarantees that problems will be overcome. In the next section we shall encounter an example where specialists deliberately avoided the term "race" only to find themselves accused of disguising an interest in race behind a new term.

across soils of different acidity. Clearly, whatever is there in the seeds will not be expressed phenotypically in every environment, and may be expressed differently in different environments. And whatever is there in the environment may affect different kinds of seeds in very different ways.

The current state of knowledge on heredity and environment is clearly displayed in this example. It can be succinctly expressed: heredity matters; environment matters; and variations in both matter for the understanding of phenotypical variations in organisms. This is the case with the plants just discussed and equally with humans. Heredity is crucial in understanding many of the ways in which we resemble each other phenotypically and differ phenotypically from other species. It is also important in accounting for some phenotypical differences between individual members of the species—the characteristic differences between those of male and those of female sex, for example. And heredity, in the form of the specific characteristics of the relevant genomes and associated cellular material, may give rise to highly specific outcomes at the phenotypical level in a wide range of different environments: the female child born in a private hospital in the UK, for example, would still have been female if guided to parturition by the National Health Service; she would have stayed female whether well or badly nourished, bombarded with ultrasound or left in peace. And she would have probably remained female thereafter in whatever environment she lived in. Even so, environment has to be allowed for: a child may remain female despite environmental variations, but even very modest environmental variation, in the normal range as it were, is known to affect weight and height and innumerable other characteristics. And of course it is easy to vary the environment sufficiently radically to change the inherited genetic information itself, whether indiscriminately or in a controlled and systematic way.

This "interactionist" account, however, has proved not to be enough to satisfy everybody's curiosity, and a great deal of work has been done in an effort to answer the further question of *how much* of the trait variation between individual human beings can be attributed to heredity, H—or more precisely, to variation in H—and how much to environment, E—or rather to variation in E. On the face of it, this seems a perfectly reasonable question. Thus, we could ask re our simple example: how much of the variation in the redness of the garden flowers was due to variation in H and how much was due to variation in E? We could check, and perhaps answer that all the observed variation was between flowers from different seed packets—strongly suggesting that none was due to variation in E and hence all was due to variation in H. With the observed germination rates, on the other hand, which packet a seed came from might prove to have been irrelevant, suggesting that all variation was the product of variation in E.

Nonetheless, we need to remember that the variation at issue here is only that which was observed with *those* seeds, in *that* garden, i.e., with those particular populations of plants, in that particular "overall" environment. And similarly, insofar as the key questions at the forefront of the H/E debates are legitimate ones, they must be understood as questions about trait variations within and among specific human populations as they exist in specific "overall" environments.[4] Crucially, these are questions that cannot sensibly be asked of individuals and individual characteristics, for they concern not the provenance of traits themselves but of variations in traits. It makes no sense to ask about the relative importance of H and E in shaping an individual nose or accounting for a particular IQ score: one can only answer that both were essential; that without the zygotic genome(s) from which the individual developed and without the oxygen she breathed from birth, for example, neither nose, nor score, nor individual would have existed. But this trite answer is hard to reconcile with some of the folk thinking that used to surround inheritance and currently helps to sustain astrological genetics—thinking that focuses on individual ancestry as a source of pride and standing and on an imagined "strength of inheritance."

Population genetics, the study of inheritance as it occurs within populations, has not sought to assess "strength of inheritance" at all, for the best of reasons as we shall see, and has concerned itself instead with *heritability*.[5] The heritability of a trait, in a given population and a given environment, is defined as that proportion of the observed variance in the

4 In the context of the H/E debates, which have been obsessed with the comparison of different groups or populations of humans, blacks with whites, males with females, and so forth, it is of particular importance to remember that these questions apply not only to variation within *specific* populations, but also variation among specific *populations*.

5 Sufficient critical leverage on misconceived versions of hereditarianism is easily gained by pointing to their tendency to misuse the notion of heritability. There is no need for us to challenge the credentials of the notion itself, although this could be done at the cost of a more technical discussion. Among the many problems such an account would encounter, the interaction of hereditary and environmental variation is the one that gives rise to the most controversy. There is considerable debate among specialists about how this interaction might be best represented in their mathematical models.

trait that can be explained by reference to variation in H. It is assumed that another part of the trait's variance will be the product of variations in E, and that variations in H and E will together somehow account for all the observed variation in the trait. Thus, an estimate of heritability is an attempt to measure the extent to which a phenotypical variation among the individuals in a given population is related to variation in their inheritance. It is not a measure of the extent to which a trait, or even the variation in a trait, is inherited or innate.[6] Still less is heritability an indication of how hard it is to change the expression or distribution of a trait in a population by changing the environment. For example, it is now quite uncontroversial that measurements of the heritability of IQ have no bearing on the question of how much IQ might be increased by changing the nature of schools or the behavior of teachers.

Measures of heritability are not inferior substitutes for measures of the strength or importance of inheritance. It is not that these latter quantities are of more interest than heritability but more difficult to measure; it is that they are not measurable quantities at all. Phenotypes and their traits are always the products of heredity *and* environment. Both will figure as necessary causes of whatever is to be accounted for at the level of the phenotype. Intuitively and informally it is reasonable to assume that variations in traits that are unaffected by the most frequently encountered or easily imaginable forms of environmental variation are related to genetic/genomic variations, and are "inherited" in the sense that different genes/genomes are crucially implicated in the processes that bring them about: it is usually expedient and harmless to speak, for example, of biological sex or of hemophilia as "inherited." But if one values accuracy and yet still hankers after some quantitative expression of the importance of inheritance, nothing better is available than heritability, which relates variation in a phenotypical trait, in a given population in a given overall environment, to variations in H (and by implication to variations in E as well). Accordingly, those with hereditarian inclinations have been forced to express themselves in the framework of the genetics of populations, and to employ a notion of heritability that does not always suit their purposes as well as they would wish.[7]

6 Philosophers have given considerable attention to the question of what it is for a trait to be innate (see, e.g., Ariew, 1996).

7 Our concern here is not with the environmental side of the H/E debates. If it had been, we should have also described how the setting of misunderstandings about heritability to rights may also weaken some environment-leaning explanations.

Of course, hereditarians also face the problem of actually measuring heritability, which is not an easy task, particularly where human psychological traits like intelligence are concerned. It is salutary to remember that currently accepted estimates of the heritability of IQ have been generated indirectly, without any knowledge of how genetic/genomic states of affairs are actually linked to intelligence. It has simply been assumed that variations in large numbers of unknown genes help to account for the normal distribution of IQ scores in the population, and the contribution made by these variations is inferred by looking at situations where they are believed to be absent. The favored method for measuring the heritability of IQ is to look at sibs and other close kin, who are more alike genetically than people in general are, and to compare variation in IQ among them with that among unrelated individuals.

The existence of monozygotic twins is a particular boon to those who would study heritability, for it provides them with a remarkably effective probe.[8] The twins in such pairs have generally been reckoned near enough identical genetically and genomically so that none of their phenotypical variation can be the consequence of variation in H. And it is precisely because of this that the effects of variation in H have been inferred from the study of twins of this sort. The overall variation in a population trait like IQ is a function of variations in both H and E. Its variation between identical twins in the population is presumed to be a function of E alone. By subtracting the latter from the former (the actual operation employed is a little more complicated), a measure of heritability may be obtained. This, roughly speaking, is how the heritability of IQ in various populations in the US and the UK has been estimated, with a wide range of results ranging up to as high as 80%.[9]

8 Given the relative rarity of twins, other blood relatives are also employed to make estimates of heritability. The correct way of proceeding here, however, is far more difficult to describe, and indeed all but the most specialized and technical discussions, including practically all of those in popular science texts, offer grotesquely inadequate and seriously misleading accounts of what is involved.

9 The most significant cause for qualification here, if by no means the only one, is that female identical twins are, in effect, genetic/genomic mosaics (see also chapter 2, pp. 66–67). Significant phenotypical variation is known to be related to this mosaicism (Migeon, 2007). Moreover, the silencing mechanisms that produce it are increasingly suspected of acting on chromosome segments as well as whole X chromosomes, which implies their ability to diminish the required "identity" of "identical" twins of both sexes. Whatever else, differences linked to mosaicism, even if they are distinguished from "inherited" genetic differences

An understanding of the role of twins in measuring heritability is relevant to the discussion in the next section. By now readers may have wearied of our account of the notion, with its repeated references to variation in traits where others talk simply of traits, and its insistent contrasting of heritability with strength of inheritance. But talk of the high heritability of a trait *sounds like* talk of things strongly controlled by inheritance and resistant to environmental variation, and it is often heard as such. And this, along with the associated errors we have been trying to avoid, continues to have baleful consequences, especially where the behavioral and psychological traits of human beings are being spoken of. These are errors that have been exposed and warned against again and again in both general and specialized literatures, but they stubbornly persist. It is as if people wish to make them, and perhaps they do; for if "highly heritable" is read incorrectly as "inherited," then assertions of high heritability nicely reinforce astrological genetics and offer an excuse for presuming the irrelevance of "the environment." And the resulting explanations of traits in terms of the power of heredity can be used to do political work for which authentic accounts of heritability are themselves unsuited.

Astrological genomics

Research in genomics and genetics continues to try to account for variations in human psychological and behavioral traits, and the fear remains that researchers will fail to distance themselves sufficiently from astrological genetics and the worst kinds of hereditarian "gene-for" discourse. It is a justified fear, even though older forms of misbegotten hereditarianism have now lost much of their credibility and are no longer used to press for state-sponsored eugenic social policies. The market for hereditarian accounts of human difference at the collective level seems to be in secular decline in Europe and probably in the USA as well, and certainly the amount of explicit opposition and revulsion elicited by such accounts is greater than it has ever been. But even if these ever more individualistic societies are increasingly disinclined to build hierarchies on group differences, especially the "racial" differences, around which the H/E debates previously played, the demand for such accounts

---

by eccentric definitions, will still perturb twin studies to the extent that they exist and ought to be allowed for when interpreting their results. However, as twin variability is *subtracted*, the existing results could be interpreted as lower-limit estimates of heritability.

has by no means disappeared altogether.[10] And at the same time, we are witnessing the rise of new eugenic ideologies, thoroughly in tune with modern individualism, and the corresponding rise of new forms of hereditarian discourse, often no less suspect than what went before even if less obviously malign, in which the individual has pride of place rather than the group or the population.

A major project the reception of which highlights changing attitudes toward group differences was the Human Genome Diversity Project (HGDP), the plans for which were drawn up only very shortly after the start of the Human Genome Project itself (Reardon, 2005). Unlike the HGP, the HGDP is generally reckoned a failure, and although the kind of research it contemplated—studying DNA from cell lines created from individuals in a selection of different human populations—is still taking place today, it is mostly under rubrics that, in contrast to the HGDP, are careful not to give undue prominence to their interest in group differences. Not that the HGDP proposal proclaimed its own interest to be primarily in group differences. On the contrary, the HGDP was rationalized mainly by stressing its relevance to an understanding of the evolutionary history of the human species. And at least some of its authors expected its findings to refute many of the claims about human difference advanced in racist and crudely hereditarian polemics. But these authors nonetheless set out their technical ambitions in a framework easily confused with that of everyday discourses of racial and ethnic difference. And the use of this framework, together with the insensitivity apparent in how it was deployed, was a significant factor, according to Reardon, in the failure of the project. Although accounts of the proposed research avoided talk of races and spoke of its authors' interest in populations, "population" was a concept easily seen as a surrogate for "race" and one that signaled the danger of race being "let in by the back door," as Reardon puts it. Nor did it help that a plea for urgent action was made, lest some of the populations seen as crucial to the project should vanish before they could be studied. This plea was not uniformly well received by the populations referred to, who had no desire to vanish any time soon. The planners found themselves criticized as exploitative, racist, and careless of the rights and dignity of indigenous peoples, and their project

10 If we look beyond the US and Europe, things become much more complicated. Notions of race need to be applied with great care in some societies, and not unthinkingly equated with notions of color, which although also discriminatory may actually be strongly individuating, as seems to be the case in Brazil, for example. See Banton (1983) for a detailed comparative account.

was denounced as the "Vampire Project." Subsequent opposition from both lay and expert sources proved sufficient to deny the project both the cooperation and the resources required for it to succeed.

Whether this failure of an apparently well-intentioned attempt to study human genetic diversity is something in which to take unalloyed pleasure is moot, but it does illustrate the changed political context surrounding research on genetic/genomic difference and the increased hostility of powerful interests to anything liable to naturalize politically problematic distinctions among human groups. And if the support of research and acceptance of its findings are going to be conditioned by broad political considerations of this sort, then at least those that operated here are more defensible than some.[11] However, while the failure of the HGDP nicely exemplifies the declining support for, indeed the declining willingness to tolerate, studies that map genetic/genomic differences among different human groups or populations, it is perhaps more interesting to look at the kind of work for which support is growing. Typically, this is much more individualistic and "psychological" in style and avoids making comparisons among social groups. Even so, the lapses to which hereditarianism has traditionally been prone remain in evidence in much of this newer research. And the genomic data and techniques that it increasingly employs are often assimilated into traditional genetic frameworks and interpreted more widely in standard astrological terms. Certainly there are cases where it is tempting to speak of an astrological genomics emerging alongside astrological genetics, with the two together being used to give new twists to what remain for the most part traditional "hereditarian" forms of argument.

The example we have chosen to discuss here is a recent study of "genetic influences on female infidelity" (Cherkas et al., 2004), the results of which were summarized in typical astrological fashion in *The Times* under the headline "One in five women strays but maybe she can't resist— it's in her genes." Note that the study was not an attempt to identify or account for differences or variations, whether at the group or individual level. It describes itself as a study of "genetic influence" on a specific human behavioral trait. Elsewhere it speaks of having demonstrated a genetic basis for infidelity (p. 654). But its key result, the result that

11 In her history of the HGDP Reardon (2005) claims that such conditioning is unavoidable: we need to accept that the natural and moral orders are invariably coproduced and cannot be disentangled from each other. Whatever else, the separation of fact and value still assumed in most Anglophone philosophy of science does need to be systematically questioned (see Kincaid et al., 2007).

demonstrates genetic influence, is that infidelity is significantly heritable. We have stressed that heritability is only a measure of the link between trait variation in a population and inherited genetic variation therein, so here, on the face of it, is a counterexample implying that our account might be inadequate. Here is a case where the significant heritability of a trait seems to be cited precisely in order to show that genetic influence partly accounts for the existence of the trait itself.

Clearly, we need to look at the claims of the study in a little more detail. Let us begin by looking at how it measures heritability. The measure is made in an entirely standard way. The researchers make use of one of the large "twin banks" that now facilitate work of this sort. They find that with regard to sexual infidelity the behavior of female monozygotic twins is more alike, twin to twin, than is the behavior of all-female dizygotic twins.[12] Loosely put, if your identical (MZ) twin has been unfaithful, then you are more likely to be unfaithful yourself than you would have been if your unfaithful twin had been a fraternal (DZ) twin. Those with least genetic variation vary least in infidelity. And by focusing on this difference between kinds of twin it is estimated that the heritability of this trait, infidelity, is 40%. Thus, if the study is to be relied upon, infidelity here is indeed significantly heritable: genetic/ genomic variation appears to account for a significant part of the variation in the incidence of the trait. And although we need to remain alert to possible failings of the kinds commonly encountered in studies of this sort,[13] this one does seem to be as reliable as most.

So far, then, so good. There is no clash between what we have said of heritability and how the authors of this study measure it. Their calculations do involve differences and variations, as they should. However, the authors are not primarily concerned with these differences and immedi-

---

12 Unconsidered problems raised by mosaicism arise with female twins, although this is not a specific failing of this study. And as far as behavior is concerned, it is *reported* behavior that is studied, and most of the relevant data was gathered by postal questionnaire. There are of course well-known biases in the reporting of behavior, but we are not offering a methodological critique of this work, and indeed it is a technically impressive study by a team with considerable experience in the calculation of the heritability of human phenotypical traits.

13 The authors certainly merit criticism when they project their heritability estimate onto other environments, including long past and arguably entirely different environments, without any further justification, regardless of the fact that heritability measurements cannot reputably be extended in this fashion. Here indeed is an instance of a ubiquitous form of misuse of this kind of finding.

ately background them once an estimate of heritability has been made. Thankfully, perhaps, they say nothing of differences between groups or kinds of women, and have no inclination to use their results to blame or stigmatize any of their fellow human beings. Their heritability measure serves mainly as a sign that further genomic research is worthwhile. When the significant heritability of an interesting trait is established today, alleles of genes or sequence variants in strings of genomic coding DNA are assumed to be involved in its expression, and research deploying an impressive battery of genomic techniques is then undertaken to identify the bits of genomes involved. A number of (in silico) genomic techniques were indeed deployed in this study once the significant heritability of infidelity became apparent, although little needs to be said of the details of this work here.[14] As it happens, no positive results were produced by this part of the research project. No doubt *The Times* will let us know if and when the genes for female infidelity are eventually discovered.

That infidelity, considered as a human phenotypical trait, has in some sense a genetic basis is hardly worth stating: without genes and genomes there would be no such traits and no humans to manifest them. That variation in that trait and/or its expression seems likely to be linked to some extent to genetic variation might be thought a little more interesting. Certainly, at some point in our history it might have interested moralists or those concerned to defend certain kinds of political and institutional arrangements constraining on women. But while this second formulation, and not the first, is clearly the finding our researchers "really meant" to report, and certainly the only one of the two that their work comes anywhere close to justifying, it is equally clear that they themselves had no direct interest in its possible moral implications. Why then did they regard this finding as interesting?

14 Data on the genomes of most of the DZ twins were examined extensively but at low resolution, seeking locations where twins alike in behavior were more alike than average in genomic inheritance. This "linkage" study was made across the entire genome and included all the chromosomes, in an impressive display of the technological resources now available even for modest studies such as this one. And had locations such as were sought actually been found, further work could then have been done to identify "genes," although in fact no such locations were identified. In addition, some of the genome data were examined at higher resolution, to check the hypothesis that alleles of a specific gene, "the AVPRIA gene," might explain variability in infidelity, but once more results were negative, and no evidence that this gene had a role was found.

Part of the answer may be that such a finding about infidelity was thought likely to be of general moral interest, and hence something that might attract attention, increase the wider visibility of the work and increase the chances of support being attracted to it. But the very act of describing the finding as a finding about infidelity involves doing violence to language for less than scientifically reputable reasons. The study defines and identifies infidelity as engaging in sexual activity with another while living in a sexual relationship with a partner, but in everyday discourse infidelity exists only where partners expect each other to forgo sexual activity with third parties and a morally sanctionable breach of faith is involved. Thus, what could be mere behavior as far as the study is concerned is described as the morally sanctionable activity of infidelity as far as external audiences are concerned, and a questionable connection with the moral sentiments of these audiences is created.[15] Of course, the gratuitous incorporation of terms from everyday moral discourse into scientific literature in this way, while widely frowned upon, is far from uncommon, and indeed in biology such discourse is even projected upon the animal world, and used metaphorically to make sense of all kinds of nonhuman behavior. Thus, while "infidelity" is a dubious term with which to refer to human sexual behavior where no expectation of fidelity is involved, its use as a means of describing the sexual behavior of other creatures is still more problematic. Yet this kind of talk is widespread, and indeed Cherkas et al. themselves happily refer to "unfaithful" female shorebirds (p. 654).[16]

Infidelity is just one of many morally salient behavioral traits that researchers tell us there are genes for, or "a genetic basis for" as the current study puts it (p. 654), in the expectation that our interest will be strongly engaged. Homosexuality is another such, although this is arguably more an inclination than a behavioral trait, as the Anglican Church keeps reminding us. Crime is yet another example, albeit one where the

15 Cherkas et al. could have defined infidelity as morally deviant behavior of course, but then they would have found it difficult to apply the term to other creatures or to extend it as they do to the human communities that evolutionary psychologists like to speak of.

16 We can at least acquit the present authors of the worst sin of all encountered in this context, which is first to apply inappropriate morally laden language to the "genetically determined" behavior of nonhuman creatures, as when male ducks are described as enthusiastic rapists, and then to take this "rape" behavior of ducks as evidence that the human behavior properly referred to as rape is "natural," since rape is genetically determined (Dupré, 2007).

idea of behavioral definition is so ludicrous that references to "a gene for crime" are now more commonly found in jokes and parodies of hereditarianism than in the real thing. Not that astrologically inclined researchers are at all discouraged by ridicule here: aware that crime is a touch too culturally loaded, they speak instead of genes for aggression, violence, lack of control, all of which are but a nod and a wink away. It is interesting to ask here what we would do if and when we gained the ability to screen populations to detect the genes or alleles "for" these things. After all, a genomic marker that correlates very strongly indeed with crime was discovered by researchers long ago, and use of the most modern and sophisticated techniques would merely confirm its position way out in front at the head of any list. But we appear to lack the inclination to direct precautionary measures specifically against those who have a Y chromosome.

One of the most interesting things about discussions of heritable human behavioral traits is how often they echo the very strongly individualistic discourse currently prominent in everyday moral and ethical debates. Again, this is nicely exemplified by Cherkas et al. They are interested in infidelity solely as a heritable trait that may or may not be manifest in individual women. They take no interest at all in whether it is more prevalent, or more heritable, among Jewish rather than Christian women, or white rather than black. Neither the focus on group difference characteristic of the old eugenics nor the associated collectivist morality is in evidence. And while it would be incorrect to claim that they reject collectivist moralizing in favor of individualistic moralizing, their results are certainly presented in a way that highlights their possible relevance to current individualistic moral concerns.

Does it really matter if human behavior is described in a slightly inaccurate way in this work, or if its findings are "sexed-up" just a little in order to attract more attention? And does an individualistic moral orientation in the work deserve to count as a cause for worry, given that our whole moral order is now a strongly individualistic one? These are reasonable questions, and timely ones, given that the case against the traditional view that moral neutrality is essential in science is constantly gaining ground and becoming increasingly difficult to counter. Nonetheless, so long as the traditional view remains widely held, as it still is both among the general public and in science, those who adopt more morally committed ways of reporting their findings risk being misunderstood, given of course that misunderstanding is not what they actually hope for. Sexed-up descriptions conditioned by moral perspectives are liable to be misinterpreted on the traditional view as simple statements of fact, and explanations favored on moral grounds as the best or even the only explanations.

We have already noted how the newspaper headline interpreted the results of the infidelity research: "maybe she can't resist—it's in her genes." The "genetic influence on infidelity" referred to by the authors, problematic enough in itself, now becomes the irresistible power of genetic determination, capable of overriding the will and fixing the fate of the individual. This is the standard astrological genetics of the media. The researchers themselves make the connection slightly differently: women are not made out as the victims of their irresistible genes; they are said to choose the strategy of infidelity because it makes sense in evolutionary terms (p. 654). Their behavior gives "evidential support" to the theories of evolutionary psychology, which claim that individuals are disposed to choose whatever actions maximize the chances of their genes surviving and multiplying.[17] Of course, we are not supposed to imagine that people perform the requisite calculations here and make conscious free choices. Choice is forced by an evolved psychology. Evolutionary psychologists love to speak metaphorically about the differential survival of (the genes of) individuals with different heritable behavioral tendencies. The behavior chosen is, in the end, held to be caused by the genes that nature has selected, but not a shred of evidence is offered for this contention.

The research results in this case show infidelity to be significantly heritable, and thereby offer evidential support to the theories of evolutionary psychology, or so we are told; but there are other theories equally consistent with these results, and other equally plausible accounts of how the connection between genes and behavior might run. Consider, for example, that in our present environment men might conceivably be particularly strongly attracted to women with specific heritable traits, to those with 0.7 waist-to-hip ratios, for example, more than to those with 1.0, or to those with blue eyes and blond hair rather than to brown-eyed brunettes.[18] If this is so, then the MZ twins of those females who evoke the male urge to mate especially strongly will also be likely to evoke

17 Strictly speaking the claim is that people have psychological mechanisms that dispose them to reproduce the kinds of behavior that would have maximized the chances of survival of their genes in an "environment of evolutionary adaptation" generally supposed to be the last million or two years.

18 Evolutionary psychologists like to talk about waist-to-hip ratios. The doyen of the relevant research is Singh (1993). But there are recent studies that suggest that the 0.7 waist-to-hip ratio is not recognized as desirable at all in some socie-

that urge especially strongly, whereas the DZ twins of similarly exceptional females will be less likely to do so. This may result in infidelity being significantly heritable even if the relevant behavioral propensities in all
the female twins are uniform, simply because males tend to treat two individuals who are MZ twins more similarly than they treat two DZ twins.

Notice that a "hereditarian" explanation of the variation in female
behavior is still being offered here, even though there is no implication
that female behavioral variation is under direct genetic influence or control. If anyone wanted to speak of the genetic control of behavior here,
then it would be a matter of *male* behavior being controlled by *female*
genes: "maybe he can't resist—it's in her genes" is perhaps what the
newspaper should have said. Instead, an individualistic moral frame and
an astrological genomics illicitly encourage the assumption that the individual is controlled by her own horoscope. It is an assumption further
reinforced in this case by the strongly entrenched prejudices that inform
our attributions of weakness of will.[19] When the newspaper headline
implies that it is females and not males who are under genetic control, it
is not scientific findings that should be looked to in order to understand
why: we should look instead to our dominant social stereotypes.

It is most important to be aware that even if the high heritability of a
behavioral variant suggests to some that there is genomic/genetic control
at work, it implies nothing about how such control is operating. When
a trait is found to be highly heritable, there is a tendency to assume not
only that genetic control exists but also that the relevant causal path lies
entirely within the individual organism itself. But insofar as it is indicative of such a connection at all, significant heritability indicates not a
chain of genetic causation confined within the bounds of an individual
organism but a connection of some sort or other, in the given environment, between genomic variant and trait. If predatory males seek to lure
only blond blue-eyed females into infidelity, their preference for these
highly heritable characteristics could well be something peculiar to our
current environment, a matter of fad or fashion. But the consequence of
this preference would not be a spurious measure of the heritability of female infidelity distorted by environmental factors: it would be a *genuine*

---

ties far removed from dominant Hollywood-driven ideals of female beauty (Yu
and Shepard, 1998).

19 A number of social psychological theories compete to explain why we attribute will power to the socially powerful and weakness of will to the socially
less powerful, but that we do so seems to be agreed. See Barnes, 2000, chap. 3.

measure of that heritability in an environment with those fads and fashions, consequent upon a set of causal connections passing through that environment. The point is an important one; for many or most of the causal chains that lead from genomes to behavior run through the social and cultural environment. Blond-blue-eyed may possibly have been highly heritable even in the Pleistocene, but whether gentlemen preferred blonds to the same extent so very long ago is doubtful.

Such is the apparent enthusiasm of Cherkas et al. for evolutionary psychology that they not only ignore alternative theoretical possibilities, but even constrain their own ideas unduly to make a connection with it. Here is how they make the crucial link:

> The fact that psychosocial traits such as . . . infidelity . . . have a heritable component, lends support to evolutionary psychologists' theories . . . If female infidelity [is] . . . under considerable genetic influence as this study demonstrates, the logical conclusion is that these behaviours persist because they have been evolutionarily advantageous for women. (p. 654)

Let us look at this "logical conclusion." Suppose that some people are more resistant to smallpox than others and that this trait is significantly heritable. It is tempting to deduce that above-average smallpox resistance must be evolutionarily advantageous. Why otherwise would it have persisted? Note, however, that if above-average smallpox resistance is an evolutionarily advantageous trait, then so by the same argument must be below-average smallpox resistance, difficult though it is to imagine any evolutionary advantage in that. Why otherwise would it have persisted? Of course, there is an immediately obvious way of understanding its persistence that has nothing to do with evolutionary advantage but a lot to do with logic. If higher-than-average resistance has persisted then so too, ipso facto, has lower-than-average resistance: the persistence of the one entails the persistence of the other. What has actually persisted is a distribution of levels of resistance encompassing both above- and below-average levels. Again, we find ourselves reminded of the need to think of heritability in relation to trait variation in a population, not as the extent of involvement of heredity in the production of the trait itself.

Let us now return to infidelity. This trait is said to be 40% heritable. But the measurement was possible only because variation existed among the women studied, with some reporting infidelity and others fidelity. And given the way it is defined, we can confidently say that if infidelity is significantly heritable then so ipso facto is fidelity. The researchers could well have reported a genetic influence on fidelity. A *Times* headline could have speculated: "Maybe she couldn't help but resist . . . ." And we could have faced the logical conclusion that fidelity is evolutionarily advantageous for women. It is indeed an interesting question why the research was presented as an investigation of infidelity rather than fidelity. Portentous conclusions could be drawn here about our current cultural habits and prejudices, but perhaps we should not make too much of what might perhaps have been something close to a divertissement for a hardworking team of researchers.

Ironically perhaps, there are some cases where a behavioral disposition and its opposite may both be evolutionarily advantageous at once, not as a matter of logic but as a matter of fact, and Cherkas et al. (2004) actually remind us of why this could be a possibility here. They describe how infidelity may have advantages in an environment where fidelity also exists (p. 655). And related arguments can easily be invented for the advantages of fidelity in an environment wherein infidelity exists. Evidently, both strategies (as the researchers describe them) can be advantageous within the same population, which suggests that different genetic influences generative of both could conceivably be evolutionarily advantageous simultaneously. The reason for this is profoundly interesting. It is that every organism in a population figures among the constituents of what other organisms encounter as their environment. The organisms, in others words, cannot be considered as independent individual units and need to be understood holistically and systemically as the interdependent constituents of the population. And the measured heritability of their behaviors or behavioral dispositions can be that heritability only in "an environment" partly constituted by those same behaviors.[20]

Beyond astrology

Cherkas et al. (2004) offer us no significant evidential support for the conjecture that some women are more strongly inclined toward infidelity than others because of genomic differences. They link their findings to

---

20 This very striking further illustration of the inadequacy of individualistic ways of thinking in this context could be expanded into a critique of the heredity/environment distinction itself and hence the coherence of the very notion of heritability. Although we do not discuss them here, problematic individualistic assumptions are actually built right into the models on the basis of which heritability is calculated as a quantitative measure.

this conjecture through flawed forms of reasoning that nicely illustrate how easily the individualism so very widespread in this context can lead inference astray. Of course, in criticizing this reasoning we cannot claim to have shown that the conclusions it incorrectly arrives at are false. It could still be the case, for example, that there are genomic differences somewhere that do correlate strongly with the extent of the enthusiasm of different women for infidelity.[21] There is no way of disproving a very general claim such as this. And indeed it would be strange if a genomically diverse group like "women" was not behaviorally diverse in some respects in consequence of that genomic diversity. It is very occasionally claimed that genomic variation has no relevance to an understanding of differences in human behavior, and such claims, clearly inspired by the fear that distasteful "moral" conclusions will be drawn from alleged genomic differences, have value in encouraging skepticism about many of the dubious empirical findings that currently abound in this area. But such claims can also distract attention from the vital question of why the empirical connections they deny the existence of should be regarded as morally significant in any case. Blanket denials of any links between genomic and racial differences, for example, may create the impression that if only there were such links we could legitimately get out our machine guns. Sometimes it can be better to celebrate the moral unity of humankind without offering hostages to the fortune of empirical enquiry.

Be that as it may, the new human genetics and genomics have to recognize the high level of interest in the wider implications of their work and that, if only by accident of history, some audiences look to them as a source of hereditarian explanations of variations in human behavior. Outside constituencies remain interested in two sorts of variation here: variation in "morally salient" propensities, in homosexuality, for example, or criminality, or mental illness; and variation in status-conferring capacities and powers, where IQ is by far the most familiar example. These constituencies still want to be told how far variations of this sort are the result of genetic inheritance, but something with which to rationalize different attitudes and ways of acting toward different human beings is what is sought, not detailed understanding. The demand is for simplified accounts of complex and/or equivocal findings, and hence for an astrological genetics and talk of "genes for." Indeed, the great attraction of an astrological genetics of human behavior as an everyday cul-

21 Henceforth the tag "in a given population, in a given environment" will be taken as a given.

tural resource is surely the ease with which it can be used to rationalize the attitudes and institutional structures that order the different ways in which we treat people.

The new genetics and genomics will inevitably be used, just as classical genetics was used, as a resource with which to rationalize social practices and institutional arrangements. Any vision of the future that failed to recognize this would be misconceived. We need to be aware of the cultural uses to which classical genetics were put, displeasing though they frequently were, in order to anticipate what may in due course happen with the new sciences. But although genetics, and human biology generally, have repeatedly been employed to rationalize our treatment of our fellow human beings, what rationalization actually involves and why it is resorted to remain unresolved problems demanding continual reflection. One very important point about all such forms of rationalization is that they are invariably contestable and lack compelling power, whether as rational argument or persuasive rhetoric. This point and its significance are best conveyed via examples.

Consider recent claims that a gene for homosexuality has been found. The relevant evidence derives from twin studies and is problematic to say the least, although this need not concern us here. What is interesting is how this "discovery" intensified the ongoing debate about what a genetic basis for homosexuality would imply at the moral and practical level, and whether self-described homosexuals ought to question, or alternatively welcome, reports of its existence. Some people were worried that "gene for" discourse would be used to reinforce and further justify the view that homosexuality was a pathology akin to a physical illness, associated with an abnormal gene or allele. And indeed genetic and other biological theories of homosexuality have long been deployed by those intent on persecuting homosexuals, in the evident belief that they identify the condition as unnatural, abnormal, or pathological. But there were many who took the opposite view, one that now seems to be increasingly favored, according to which a genetic account implies that homosexuality is a natural state or propensity, and exposes moral condemnation as unjustified. The relevance of this example here, of course, is as a reminder that scientific findings, considered in isolation, imply nothing beyond themselves. If homosexuality is found to be significantly heritable, or under genetic influence, or even (heaven forbid) the product of a "homosexuality gene," that in itself implies nothing about whether homosexuality is natural or unnatural, justifiable or unjustifiable, acceptable or reprehensible. Any inference from the finding to a conclusion of this sort will be contestable, and the problem will remain of how such a finding can rationalize a given mode of treatment or moral orientation toward homosexuals.

As it is with propensities, so is it also with powers and capacities. Intelligence is widely accepted, at least by those willing to concede that there is such a capacity, as being highly heritable. And variation in intelligence, as measured by IQ tests and represented in normal distributions of IQ, is sometimes thought to justify discriminatory educational practices of various kinds, just because of its high heritability. But although the relationship between knowledge and practice here is widely perceived as obvious, on closer examination it is exceedingly strange and difficult to understand. Recall how the variation of IQ in a population is conventionally represented as a normal distribution around an average IQ of 100. This distribution and its high heritability have been widely cited in rationalizing the current tendency to concentrate educational provision upon students of above average IQ—students "more able to benefit" as is sometimes said. But we do not extend this argument to those of very low IQ and offer them correspondingly fewer resources. When scores reach 75 or less there is typically a switch of policy; extra educational resources are provided for the groups involved, and we actually tend to regard that policy as justified precisely by the low IQs of its beneficiaries. IQ scores do not constitute a single bell-curve or normal distribution. They constitute a distribution with two peaks, one conventionally set around 100, and another, very much smaller but still more highly heritable, at around 70. At and above the higher peak it seems rational to many people to concentrate resources on the more intelligent. But approaching the lower peak it seems rational to many of the same people to concentrate resources on the less intelligent and to argue that this is justified in terms of their "special educational needs." As in the previous example, it is not the arguments themselves that are interesting here but the evident conflict between the one and the other and the fact that many people nonetheless appear content to accept both of them. There could hardly be a more striking illustration of the less-than-compelling character of rationalizations and justifications.

As we have said, genomic studies of human behavior are bound to be used to rationalize policies and practices, and findings suggesting that genomic differences are linked to behavioral differences are going to be cited to rationalize different ways of treating the people involved. Even though such a rationalization never has the power to compel and is never more than one of many possible stories linking knowledge to action, it is clear empirically that the fashioning of links of this sort is regarded as of great importance in debates about what constitutes sound policy and

moral action, especially where the links are to the authoritative empirical knowledge of the sciences.[22] At the same time, however, the fact that
rationalizations are never compelling helps us to understand the intermi-
nable nature of so many of the debates in which they feature, and how
opposed factions in our morally diverse and politically divided societies
are never at a loss for a response to the "scientific" arguments of their
opponents. Protagonists are free to claim, in these debates, that the science cited by the enemy is wrong, or is being misunderstood and incorrectly employed, but while they often do one or the other of these things,
they have no need to. For they may always claim that what the science
is said to imply for the political or moral issue being debated is not what
the science "really" implies for that issue. What has happened in the
case of the heritability of IQ and educational policy, or with "genes for"
homosexuality and moral evaluations of that sexual orientation, may
always happen in these kinds of case. Both sides may always challenge
the way in which their opponents make the move from the realm of science to those of politics and morals.[23]

It is sometimes said that genomics is more liberal and humane than
classical genetics ever was, and devoid of the determinism that made the
earlier science the source of so many pessimistic and conservative conclusions. But if it is deeply problematic to speak of specific moral and political implications of the findings of a science, it must be still more so to
identify an inherently moral or political dimension in the science itself. We
should be content merely to refer to the predominantly reactionary uses

22 All the sciences find their knowledge cited in such problematic rationalizations. Brain scans using positron emission tomography indicate that the human brain continues to grow and develop well beyond the second decade of life,
and they were apparently taken into account by the US Supreme Court when it
ruled that capital punishment for criminal offences committed before the age of
eighteen was unconstitutional. But just what the connection is supposed to be
between the scientific findings and the legal decision is obscure. There are clearly
fascinating parallels between the moral and political use of studies of the brain
and of genetic/genomic studies. We thank Sheila Jasanoff for bringing this particular instance to our attention.

23 The insecurity of moves from science to morals may be thought of as
the analogue in differentiated societies of the insecurity of moves from "is" to
"ought" famously exposed by David Hume. Hume is now often accused of as-
suming a sharp, and by present lights naïve, dichotomy between the factual and
the normative. But whatever else, naïve Hume was not, and it is worth recalling
how his reflections on "is" and "ought" were prompted precisely by an awareness of how easy it was in practice to move from the one to the other.

and interpretations of the old genetics, and to contrast them if we must with those uses and interpretations of genomics that are presently more morally and politically progressive, often reflecting societies that have themselves become more morally and politically progressive. Primarily, it is we who have become less conservative and hierarchical, and the supposedly changing implications of our science are changes in what we are making of it more than changes in "the science itself." And in the face of whatever further moral and political changes our societies subsequently undergo, there is nothing in our current sciences to prevent their current moral and political implications being modified and reoriented appropriately. In fact, of course, change of this sort is unremitting, and new ways of drawing moral and political implications from scientific knowledge are constantly invented in response to it.

We have already mentioned one change to be expected here. As interest in the rationalization of group distinctions weakens, so interest in the rationalization of individual distinctions is becoming relatively more important. Thus, we should expect more and more reports linking genomic locations with all the many and various individual traits that psychologists theorize about, and with aspects of individual personality that vary within groups as strongly as they do among them. And we should expect to see similar work proliferating on individual pathologies, such as might facilitate and rationalize a helpful individualistic eugenics rather than a negative collectively oriented one. This in itself, however, will represent a very modest change. Links of this sort may still be made in the astrological explanatory frame and used to rationalize attitudes and policies as findings set in that frame have long been used. And researchers themselves will be tempted to continue to make use of this frame, if only because its familiarity and wide dissemination facilitates communication with large outside audiences.

Ought we to expect changes more far-reaching than this? Perhaps not, or not immediately, but we can nonetheless hope for them and encourage them. Following on from the transition to genomics, it would be good if ways were found to strip the remaining ontological authority from familiar astrological stereotypes and make everyday understandings less dependent on them. And as far as human behavior is concerned, a significant amount of unwanted historical baggage would go with them if the stereotypes could actually be discarded. How though might the relationship of genomes and behavior then be thought of, outside the specialized fields that study it? The specialists themselves have long been putting forward one possible alternative framework. The primary role of genomes as they are currently understood is their role in the production of proteins. Hence, if there is to be a genomics or a molecular genetics of behavior, it must be because complex causal chains exist leading from genomes and the associated molecular systems, through proteins and other transcription products, to cell chemistry, cell function, ontogeny and differentiation, and thence to the structure and internal functioning of the whole organism including the brain, and on to behavior.[24] Here is a scheme that might orient even everyday thinking: we may conceptualize the connections between genomes and behavior as really consisting in extended complex chains of causation and not the direct determining connections of astrological genetics.

Tracing causal chains from genomes toward the manifest phenotypical characteristics of organisms, in which are included their behavioral dispositions, can be seen as a routine extension of the familiar "bottom-up" explanatory project supported by many researchers, beginning with the study of the various "omes" and seeking to move on from an understanding of small things to an explanation of larger ones. Of course, this is itself a much-criticized scheme,[25] and it does indeed face a number of problems and difficulties, although in our view these are not fundamental ones when it is appropriately applied. Certainly, there are few on either side of the heredity/environment debates who would deny the existence of causal chains connecting genomes and behavior. That no genomes means no behavior is scarcely controversial. Equally, however, few on either side would deny how formidable is the pragmatic problem of successfully identifying and investigating these chains. Complex as they are in themselves, they are just particular subsets of connections in far larger and still more complex causal networks. Moreover, the brain, through which the chains have to pass, is an organ that is constituted by the environment in a more profound way than other bodily organs are, and whatever is mediated by the brain will be significantly affected by the specific environment, physical, social, and cultural, in which the brain continually develops. On both these counts the investigation of causal

24 We have already stressed that these causal chains do not lie entirely within the organism itself.

25 Dupré (1993) has offered an extensive critique of the reductionist methodology referred to here. However, the intent of that work was not to deny the obvious value of reductionist explanations but, in keeping with the proper sense of "critique," to explore their scope and limits. The general thesis defended was that whereas reductionist explanation is indispensable in accounting for the capacities of things, it is typically inappropriate for explaining why capacities are exercised on particular occasions.

connections between genomes and behavior, and particularly voluntary behavior, must be expected to be extraordinarily difficult to map out and study. And it is hence scarcely surprising that research on these connections has so far failed to throw much light on variation in the kinds of behavioral dispositions that have been concerning us.

If the causal connections between genomes and behavior are to be investigated directly, then in vitro laboratory research must be involved, and where the earlier parts of the causal chains are concerned, cell chemistry and cell biology will be bound to figure large. Work of this sort is worlds away from the study of behavioral traits in large human populations, but many of the problems faced by the latter also arise in relation to laboratory studies. It would be incorrect to assume that our earlier discussion of heritability applies only to the production of visible traits in human populations and not to the production of DNA transcripts and proteins in human cells. Formally speaking, it applies in both contexts, and all the objections to unqualified "hereditarian" explanations developed in the context of the familiar H/E debates also apply in both. In particular, accounts of how some molecules are implicated in the production of others in cells, like accounts of how alleles are implicated in the production of traits in human populations, ought ideally to take account of the environment.

Why then is there not a general and widely expressed concern with the role of "the environment" here? The obvious answer is that cellular environments are neither so familiar nor so interesting to wider audiences as social and cultural environments. But while this is no doubt correct, the way that the cellular environment has been studied also helps to account for this relative lack of interest. Outside audiences have been accustomed to look to genetics for accounts of human difference, and to critics for alternative accounts of that difference as the product of environmental variation. But work on what is occurring at the cellular level has methods of making environments already resistant to external disturbance and modification still more stable and uniform, and it deploys them with the aim of understanding widely occurring biological processes that may not vary significantly among human groups or even between humans and other kinds of organism. Although this kind of work does of course take account of differences and variations, it is not forced as the older genetics was to confine its explanations to them. Indeed, in practice researchers can now often forget about difference and variation without serious harm, and simply ask: "How does A cause B?" or "How does A inhibit the production of B?" on the implicit assumption that these processes occur in all "normal" cellular environments.

One result of this could well be a shift in the nature and intensity of the controversies surrounding human genetics and genomics. New debates on the nature of human nature and the problem of what it is to be human will surely continue to proliferate around them, but despite their portentous rhetoric these could well prove far less vicious and divisive than old disputes about racial and sexual difference or even more recent ones about the heritability of individual psychological traits and aspects of personality. A thoroughgoing reorientation of this sort, however, wherein the new agenda actually replaced the old, would entail the eclipse of an astrological genetics that is currently showing no signs whatsoever of imminent demise. It may indeed be that the position of astrological genetics in the public realm will be significantly weakened and the forms of causal explanation actually employed in genetics and genomics fully appreciated only when the powers of those sciences are developed still further and confront us more directly and extensively in the form of practice, whether to our benefit or otherwise. Breaks with tradition at the level of ideas are generally prompted by major changes in how things are done and how lives are lived, and not vice versa. Whether the new powers currently being engendered by genomics will come to have such significant effects on our living of our lives is one of the main questions to be addressed in the following chapter.

6

Genomics as Power

Accumulating powers

The problem of the relationship between knowledge and power in society has not always attracted the attention it now receives. A quarter of a century ago our interest in power was largely in who possessed it and why it was so unevenly spread. Concern with the distribution of political power diverted attention away from the question of what power actually consisted in and how it was generated and sustained. And a sense that hierarchies of power must somehow be maintained through deception and other less than fully reputable means created interest in the links between power and ideology but not in those between power and knowledge. The radically different situation today owes much to the influential French intellectual Michel Foucault (1977; 1980), whose account of power was constructed in diametric opposition to previously accepted perspectives.[1] Instead of hierarchical distributions of power, Foucault spoke of its all-pervasive character and its presence in every pore of social life. Instead of linking it to ideology, he spoke of its intimate connection with

[1] Several important sociological studies of the biological sciences have been framed in Foucauldian terms (Rabinow and Rose, 2003; Rose, 2007).

knowledge; indeed, he spoke of the different forms of truth engendered in or by different systems of power. No less important, however, has been our belated but now widespread recognition of the extent of the dependence of political power on scientific and technical expertise, and how much political controversy today is at the same time technical and scientific controversy.

A major consequence of subsequent reflection here has been more and more awareness of just how intimate the connection is between knowledge and power: it is now commonplace to hear the two things spoken of as inseparable, indissolubly connected, inextricably tangled together, mutually reinforcing. But the crucial move in this direction, the move that can result in a major qualitative shift in understanding, is to recognize that knowledge and power are not separate things at all and that even to speak of their intimate relationship is to perpetuate a misconception: knowledge and power are the same phenomenon understood from different perspectives. Think of a group of humans possessed of the inheritance of knowledge, skill or know-how, tools and artifacts, social relationships and patterns of organization, with which such groups are always endowed. By virtue of that inheritance the group will be able to do far more than a mere aggregate of organisms of equivalent size. Its powers, and especially its collective powers and capacities for joint action, will be vastly increased. It is actually an enormous help to the imagination in this context to speak of "powers" rather than "power," because powers are immediately and intuitively understood as being capacities and capabilities. Since any description of what a group is capable of doing will be very close to a description of what it knows, and will at very least need to incorporate much of the latter into itself, a sense that knowledge and power are one thing differently conceptualized naturally emerges, and the claim that power is knowledge no longer stands in danger of sounding like abstract theoretical speculation.

There are critics of the notion that power is capacity to act or to do things, and indeed there are interesting criticisms to be made of it. One of these points out that just what a capacity consists in and just what the actions are that it allows to be performed can never be exhaustively specified. Another is that two powers combined may permit much more to be done than two powers employed separately: two bottles in combination have the same storage capacity as the bottles employed separately, but powers are not like that. Even so, the view of power as capacity, or capability, or potential provides the basis of the only defensible extant accounts of what power consists in. And indeed it leads to a very satisfactory account of the nature of power and one that offers a good basis for an understanding of the salience of science and technology (Barnes, 1988). As carriers of knowledge the sciences are ipso facto carriers of powers; and in engendering and sustaining new knowledge they constitute new powers. It is important, however, to avoid the further inference that the sciences, or scientists themselves, *possess* the powers that they and their knowledge constitute. Possession betokens a right, closely akin to a property right, a right of use and disposal. The powers scientists constitute are typically at the disposal of nonscientists, who have the greater part of the recognized discretion in how or whether they should be exercised, just as more generally in society the knowledgeable human beings who constitute powers are not the same human beings as those who determine how they are exercised, those who in everyday discourse are somewhat misleadingly referred to as "powerful."

However obvious it may be, there is a whole host of reasons for constantly keeping in mind this distinction between the constitution and the possession of power(s). Not least, we need to be clear that as a possession power is alienable, as property is, that it may be expropriated and redirected, as US technology was by those who used it on 9-11, or redistributed through the actions of those human beings who constitute it, as many industrial and financial organizations know to their cost.[2] One of the ways that this is most salient to the sciences is that it results in the fear so often directed toward them taking two forms: there is a fear of science as the ever increasing power possessed by those whom in effect it serves, the elites and dominant institutions in society; and there is a fear of science as power per se, not an altogether unreasonable fear it might be thought, given the awesome powers that science is engendering and the lack of guarantees that all of those who come to wield those powers in future decades, not to say future centuries, will wield them well.

All the sciences engender powers and capacities as research proceeds: even the purely observational science of astronomy facilitated navigation; and geology, similarly nonexperimental, has assisted the identification of oilfields and mineral deposits. Thus, it is always possible to

---

2 The examples here deliberately conflate "physical" and "social" capacities, and indeed it is a merit of the "power as capacity" view that it makes such a distinction unnecessary. Both social organization and skill in the use of physical objects or artifacts increase capacity to do things, and both entail knowledgeability. Note as well that the borrowing of "technology" involved on 9-11 involved not just aircraft but the skills needed to fly them as well. Sadly, as we note elsewhere, "technology," originally used to speak of the technical skills and artifacts of a culture, is now all too often used to refer only to gadgets and machinery.

address the sciences in a utilitarian frame, and to attempt to evaluate them either in terms of the net benefits currently deriving from the powers they provide or in terms of the benefits there is some reasonable hope they will make available. This is the perspective on the sciences currently most favored by external audiences, and in this chapter we shall consider how it might be applied to genomics and genomic powers. First of all, however, we need to say a little about the growth of those powers.

The power to improve populations by selective breeding has long existed and frequently been exercised on plants and animals. In the nineteenth century the use of this same power on human beings came to be known as eugenics, and the term "eugenic" continues to be defined as "pertaining to race improvement by judicious mating and helping the better stock to prevail." But in the first half of the twentieth century eugenics became notorious in the form of political programs wherein "helping the better stock to prevail" involved compulsory sterilization, and eventually mass murder, with the objective of completely eliminating specific social groups. None of the powers deployed in these eugenic programs was the direct product of research in genetics, and none of the methods employed to identify the Jews, Gypsies, mentally handicapped people, and their many other targets stands to its credit either. But genetics associated itself so very closely with eugenic ideologies, both individually and institutionally, that its reputation was tarnished and external perceptions of the field were still adversely affected decades later. It became seen, as some of its practitioners had at one time wished it to be seen, as the field that could provide a scientific basis for "race improvement" through selective breeding. And the suspicion that whatever powers and techniques it was eventually able to develop were sure to prove malign was widely entertained in consequence.

At its inception, the cultural inheritance of classical genetics was the powers of plant and animal breeders, together with generally accepted patterns of thought concerning ancestry, inheritance, race, and so forth. Fifty years later it had developed a remarkable body of specialized techniques, an impressive stock of new knowledge, a characteristic theoretical perspective known, somewhat misleadingly, as "Mendelism," and an image tarnished by eugenics. One of its first tasks at the end of the Second World War was to rid itself of this last unfortunate item, and a decontamination program duly got under way and proved notably successful. Even so, the association with eugenics continued to affect perceptions in indirect ways. In particular, it encouraged a deterministic understanding of the science and a view of genes as invisible essences that accounted for visible traits. As we have already noted, classical genetics was not inherently deterministic or essentialist; Mendel's so-called laws were known to be false almost before they were known at all. Notions like mutation were central to the very structure of thinking in the field. Chromosome studies of recombination and translocation added to the evidence that inheritance could be a hit-or-miss business. But the utility of genetic knowledge in "improving the stock" by selective breeding was most readily rationalized informally by identifying genes or their alleles as the causal antecedents, in effect the sufficient determinants, of sought-after traits. And this further entrenched misunderstandings that persist today in the careless use of the language of "genes for" characteristic of astrological genetics.

The new genetics of the late twentieth century would have been impossible without the cultural resources inherited from classical genetics, but it distanced itself more and more from those origins as its attention turned to the molecular level and its repertoire of powers and techniques came to reflect this change of focus. And just as the link to eugenics had previously distorted perceptions of classical genetics, so now the link to classical genetics distorted perceptions of the new genetics. This time, however, the distortion was not so unwelcome. Genetics had managed to wash away most of its displeasing associations with old eugenics by the end of the 1970s. In particular, it had strongly alienated itself from the group-oriented character of that eugenics, and its obsession with differences between races and other "stock" of distinctive ancestry. It had even begun to rearticulate eugenic aspirations in a more acceptable individualistic guise, and was finding support for its research as relevant to the health and well-being of individual persons. In this changed context the ready intelligibility of classical schemas often outweighed any residual negative associations they still carried, and specialists did not hurry to cast aside the veil of old knowledge and old stereotypes, even though this both concealed much of the potential of their work and helped to perpetuate a misconceived genetic determinism.

The early 1970s saw the advent of recombinant DNA technology and the subsequent abandonment of any residual doubt that human beings would acquire the power to rewrite the messages written in their own DNA. No veil could obscure the import of this development, and even conceptualized in classical idiom as the advent of genetic engineering its enormity was widely grasped: genes could now be transferred between species, including *Homo sapiens*. In effect, mating was now possible not only between the individuals in a species but between all the different species, with humans deciding precisely what each one would contribute. And this was greeted with extremes both of enthusiasm and of the horror and aversion so often elicited by miscegenation and the production of mongrel forms. Even so, something was missed by talk of genetic engineering. Today we can redescribe the advent of recombinant DNA technologies anachronistically but not incorrectly by reference to genomes and identify the use of restriction enzymes as one of the key powers that may eventually permit us to manipulate both genomes and the molecular systems in which they reside just as we choose.

Many versions of this project have been formulated. Most of us will have encountered one or another of them, and may perhaps have been repelled by their hyperbole and by the arrogant way in which the inevitability of their eventual success is asserted. Even so, the basic claims underlying the hyperbole are, we believe, likely to prove correct. Of course, we do not currently possess the powers to control and manipulate genomes at will, still less the powers to control the complex systems of which genomes are components, or the functioning of the other polymers synthesized with the aid of DNA. Existing powers amount to far more than the ability to add and subtract genes from genomes, but they will have to be elaborated, extended, supplemented, and recombined in a long period of development before coming anywhere near to justifying the kinds of vision to which we have just alluded. There is no way of intervening reliably into the process of ontogeny. The chemical links between genome and context, including key processes like DNA methylation, have not yet been adequately described let alone controlled. Even the ability to place DNA into specific locations in the genome—as opposed to inducing the genome to take up alien DNA somehow or other—is currently lacking. In a nutshell, there is a very short list of things that currently can be done, and an endless list of important things that currently cannot be done. So we need to make it clear why we are not more critical of the enthusiasts in this context, and why we believe that all the many powers we do not currently possess are kinds of powers that we could, and in all probability shall, in the long term come to possess.

For obvious reasons, proponents of new sciences and technologies are prone to make exaggerated claims for them, and particularly to misjudge the amount both of time and effort required to bring about the advances they predict. Occasionally, too, their claims become so excessive that they go way beyond any plausible imaginative projection of the current state of the art. And similarly their critics exaggerate the problems facing new fields and like to find arguments implying that they are impossible to overcome. But outside audiences are prone to underestimate the future possibilities of technologies, even if they have no interests that predispose them one way or the other, and they are particularly likely to do so at the outset of great technological developments when experience of the power that is unfolding is still limited. Genomics and genomic technology are indeed just beginning and have as yet merely scratched the surface of the possibilities extending out before them. They are visible only as a limited range of powers applied in a limited range of contexts, and our imaginations cannot easily come to terms with the ways that these individual powers are likely to grow, particularly where research and development are directed to their elaboration and recombination and actually employ the very powers themselves in order to improve and increase them.

It can be no less hard for our imaginations properly to grasp the extent of powers that already exist, not least because of the language we generally employ to refer to them. In ordinary life we tend to characterize skills and practices, artifacts and instruments, *functionally*, by reference to how they are routinely used and hence to the powers they routinely confer. This focuses attention firmly on what is established already and away from what is merely possible. Thus, we talk of screwdrivers (not paint-can openers) and of refrigerators (not heat pumps). Another striking but tragic example is box cutters. Closer to our present topic, but still firmly in the realm of everyday life, is the computer (not the typewriter or media center). Notice, however, that merely to cite these examples is to recognize that the imagination is not fully constrained by language and that our intuitive grasp of the powers we possess transcends our linguistic accounts of them. Even so, both in specialized fields like genomics and in the context of everyday life, there is a sense in which we always have more power than we know we have.

Of course, we cannot justify a radical view of the future powers of genomics merely by noting that our imaginations are not currently well equipped to anticipate them. Another line of argument is needed here. We can begin by noting that references to how things function figure prominently not just in the discourse of everyday life but in that of specialists as well. Genomics and allied biological sciences are especially prone to identify both objects and processes in terms of their functions. Whole organisms may be described in this way, as when bacteria or viruses are characterized as vectors. And molecules and fragments of molecules may also be so characterized, as enzymes for example, or functionally specified genes, or, again, as vectors, or as countless other things. Biological and chemical nomenclature meet here, at the molecular level, and the contrast between the two at this point is very striking, with chemical taxonomy mainly, although not exclusively, focused upon the structure of molecules and their constituents; and biological terms mainly, although again not exclusively, identifying them in terms of functions, and in many cases having nothing but function available with which to identify them. The boundary between the chemical and the biological here is often revealingly blurred, with the structural and functional bases of classification often being combined, as for example in m-, t-, i-, and other RNAs, wherein the prefixes are functional/biological and the terms to which they are attached are chemical, and descriptive of molecular structure.

Genomics, like molecular biology more generally, possesses a vast terminology in which objects are identified in terms of functions. But what functions are fastened upon? Broadly speaking, they are of two kinds: there are the functions that the molecular object performs in the context of the system of which it is a part, to the dynamic stability of which it contributes; and there are the functions that these objects perform for researchers. But far from being mutually exclusive, both kinds of function are available as means of identifying interesting molecular objects, and sometimes terminology moves its basis from one kind to the other. A nice example is that of the nucleases that are able to degrade DNA and function naturally in bacteria as protection against alien DNA invasion. These were indeed initially classified in terms of how they functioned naturally; but some of them came subsequently to be characterized as restriction enzymes, that is by reference to their function in the context of recombinant DNA technology, where their role was not the destruction of lengths of DNA but the production of lengths of DNA of precise size and/or sequence. The shift of functional designation here marks the point at which a natural system became a part of a technological system. It is a very common kind of shift, reflecting the extent to which natural systems have become incorporated into technological systems, or perhaps the extent to which technological systems have been created from natural biological systems. And it is a shift that could occur with any natural system: natural systems are all potential components of technological systems.

As we noted in chapter 1, living things sustain metabolic processes so sensitive to the details of molecular structure and other features of cellular micro-environments that were they expressions of human will and intention they would be accounted amazing displays of virtuosity and refined control. And increasingly we have made these processes expressions of will and intention by incorporating the living things that carry them into our technology. Living things now constitute a large part of the technology of biology as well as being its subject matter, and the important role they now play as components of the technologies of genomics and molecular genetics is clear. The traditional view that the tools and instruments deployed in technological systems are artifacts specifically designed and created by human beings should not be allowed to mislead us here. It is true that many new tools and instruments produced by physical scientists and engineers have been imported into the biological sciences and deployed to remarkable effect. Those who anticipate an unlimited control of biological systems at the molecular level through powers imported from physics and chemistry are making an understandable imaginative extension from the current state of play. But biology has also developed its own resources here and its ability to control biological systems at the molecular level depends, and will continue to depend, just as heavily on its ability to harness living things into the technological systems through which that control is exercised.

Just as the most advanced aeronautical engineers must sometimes give best to birds and be forced into admiration of the "design" of their feathers, their mode of operation, and the materials that constitute them, so the chemical engineer might feel similarly of the processes and constituents of natural cellular systems. And it is the existence of these systems that ought to serve as the clinching argument that the power and promise now widely perceived as latent in genomics and molecular cell biology could indeed be realized. The technological processes implied are there already in natural systems. The ability to transform these natural systems into technological systems, to harness and order them so that they do things at our behest as instruments and artifacts do, is already in evidence. The possibilities not yet actual in our powers to control are nonetheless actual in another sense; the task is not to invent them but only to harness them. The prospect is not of knowledge and control through the use of extrinsic powers to dismember and analyze biological systems, but of knowledge and control through use of the powers intrinsic to these very systems.

Genomics and social powers

Having acknowledged the potentially awesome powers of genomics and their future possibilities, it is time to return, with a bump, to the present, and to note how very small their current impact is, at least on the genomes of human beings. Genomics may eventually elaborate powers and procedures specifically for the reconstruction of human genomes; cloning and gene therapy are recognized as realistic future possibilities here, and there is no reason in principle why other techniques and procedures already routinely employed in the manipulation of nonhuman genomes should not be modified for use in this way. But very little at all of this kind is yet actual. And while this is a relief to some it leads others to complain of the limited extent of the benefits to human health and well-being that have so far flowed from genomics and from the vast investment in the Human Genome Project in particular. There is, however, another way in which genomic powers and procedures have found extensive application in relation to human beings, one we should count as a major cause for concern even though some would claim that major benefits are already flowing from it.

Powers need to be thought of holistically and systemically, not as so many separate, independent units. When new powers are invented they always transform the possibilities inherent in preexisting powers, and thereby redefine them. The point is well illustrated by molecular genetics and genomics, which have transformed the scope, utility, and mode of application of a whole range of existing powers, both physical and social, through the diagnostic powers they have produced. These powers are now being widely used not to transform human genomes but to generate knowledge of them, and thereby make their possessors the targets of existing powers. It is not the direct targeting of physical powers on our bodies, however, that is the major concern here, although it will become a topic of increasing importance if, for example, the aspirations of pharmacogenetics are realized and drugs are increasingly deployed selectively on the basis of genomic tests. The immediately pressing source of anxiety is our increasing vulnerability to social powers, powers to vary our rights and privileges, and our duties and obligations, powers possessed by administrators, bureaucrats, and various kinds of technical professionals, and increasingly exercised in the light of genomic knowledge.

There are now many examples of social powers being aligned and directed in this way. Indeed, most discussions of the consequences of genomics are currently of examples of this sort and no brief account can hope to convey their range and variety. Insofar as economic rights are concerned, genomic knowledge is beginning to be used to limit access to employment opportunities and to insurance, as well of course as to limit the access of firms to the marketplace through patents. In the legal context, "DNA fingerprinting" and other kinds of testing are increasingly used to identify criminals and establish their guilt or innocence (over a hundred prisoners in the US have been released from death row so far as a result of DNA testing), and these techniques are increasingly interesting not just police forces and courts of law but immigration agencies and military bureaucracies as well. In relation to basic social and political rights, the same sort of knowledge has been used by bureaucrats and administrators to establish sex (both at birth and later, as, for a time, in women's athletics), paternity, kinship, and hence rights to citizenship and participation in social life. A striking example here is that of the Lemba people discussed in chapter 4: genomic tests were used in an attempt to check their claims to be descended from one of the lost tribes of Israel and hence entitled to return to modern Israel.

It can be asked, of course, why any of this should be reckoned a source of anxiety. Is it not simply a matter of society putting genomic knowledge to good use? Have we not merely been describing some individuals being debarred from workplaces likely to do them harm, others being asked to pay fair rates of insurance, others again suffering just deserts for their crimes, and all of them being protected from the significant sources of danger represented by criminals and external enemies? As far as rights associated with gender, kinship, and citizenship are concerned, do not bureaucracies function better by making reference to genetic/genomic knowledge, and allocate these rights more justly in consequence? And have we not in any case slanted the discussion by ignoring the role of genomic diagnostics in empowering individuals and enabling them successfully to oppose institutional constraints. The Lemba in the above example were, after all, a group of people seeking rights from a reluctant bureaucracy. And much the same kinds of ancestry test involved in that case are now marketed to individual African-Americans as a means of investigating their origins and assisting them, as some say, to develop a better sense of their identity.

There is indeed a case for remaining sanguine about genomic diagnostics of this sort, but there is a powerful case the other way as well, and it is interesting to note how the case for is at its strongest when couched largely in individualistic terms as above, and how the case against grows more formidable as the limitations of an individualistic perspective are recognized. Perhaps if we had a complete and unconditional trust in our institutions and administrative systems, then we would be happy to regard any increase in what they knew of us as desirable. But the usual attitude is that they should know only as much as they need, lest they become overweening and intrusive sources of power, and that, whatever else, their dealings with one individual should not entitle them to privileged access to knowledge of another individual.[3]

3 For all that their joint work precedes the advent of the human genome projects, Nelkin and Tancredi (1989) still stand among the best guides to the issues in this context.

Genomics is a body of knowledge with implications that cannot be confined to individuals. Knowledge of the genome of one individual is also knowledge of the genomes of her ancestors, progeny, and lateral kin. One of the more controversial practices in forensic uses of genomics, for example, is to use police DNA databases and/or DNA samples mainly taken from convicted criminals to identify DNA collected at crime scenes as belonging not to one of these criminals but to one of his or her relatives. Nor is that the worst of it. It is often supposed that data about people's DNA will appear on police computers only if they have been convicted of a crime. But under current law in England, where there has been an especially strong urge to accumulate data of this sort and remarkably little anxiety about the obvious downside of doing so, one needs only to have been arrested on suspicion of a crime for a DNA sample to be taken and retained by the police, and identification data derived from it to be entered on their database. With a database compiled and used in this way, anyone at all may potentially be tracked; and as more and more data are put on file, we get closer and closer to the possibility of the universal genotyping that is an attractive idea for many law enforcement bureaucrats. And if this is reckoned a cause for anxiety, then the latent powers available to police through their retention of DNA samples are another, orders of magnitude more serious. In England, current legislation (2007) allows the police to retain these samples, and hence entire genomes, for a century, during which time all kinds of new ways of interrogating them are bound to be invented and bureaucrats are likely to have more and more "useful" ways of playing with them brought to their attention.[4]

Genomics is a potential nightmare for any society that would treat its members as separate individual units and have its bureaucracies respect their rights to privacy. Just as we cannot learn about an individual from her genome without learning about other individuals as well, so we cannot learn about one characteristic of an individual from her genome without learning about other characteristics as well. Because genomes are implicated in both the mechanisms of inheritance and in cell function in the phenotype, knowledge—say of a health-relevant "genetic ab-

4 There have been some signs that England is beginning to come to its senses on this matter and that changes in legislation could be in the offing, but even as the issues have begun at last to be debated, powerful voices are speaking out in favor of a database including all the fifty million inhabitants of the country. The situation elsewhere in the UK is different and for the most part is less illiberal.

normality"—is at the same time potential knowledge of ancestry, ethnic origin, and geographical origin as well. To put it another way, knowledge gathered to assist specialists to further the health and well-being of an individual may well also be of interest to security and immigration agencies. And in general, whenever bureaucrats, administrators, or expert professionals are in possession of enough genomic knowledge for their immediate purposes, they have more knowledge than they need and arguably more power then they should.

It is true that attempts are made to limit this excess of administrative and bureaucratic power by restricting access to knowledge and legislating to protect data on individual persons, but this is less a solution to the problem than an acknowledgment of it. A better case for permitting bureaucracies to hold and process immense quantities of genomic data, given that the data are reckoned an asset in the first place, can be made by considering how else it might be stored and processed. Imagine the possible consequences were knowledge of individual genomes to leak from bureaucratic systems into the public domain. Knowledge of this sort often results in a particular status being ascribed to the individual and her becoming the focus of particular attitudes and forms of treatment. But the nature of that treatment differs between bureaucratic contexts and ordinary social settings. In the former, genomic status markers permit administrators to manipulate and process human beings dispassionately and instrumentally, according to narrowly formulated rules, with no great regard for their special standing as human beings. But in the latter, the same status markers typically function very differently and are liable to structure whole sets of attitudes and wide ranges of behavior, because they are perceived as general indicators of honor or worth. Some of the worst problems arise here, of course, when the assigned level of honor is very low, and people are stigmatized as defective or abnormal in their very nature. The consequences of this in terms of avoidance, exclusion, emotional aversion, even universal and indiscriminate disdain, can be catastrophic for the individual, and scarcely less so for society.

Stigmatization is conventionally, but not incorrectly, regarded as a denial of full human dignity, and as such its discussion belongs in the next chapter. But it needs brief mention here to help us grasp the dilemmas raised by the proliferation of genomic knowledge. Confined to bureaucrats and administrators it enhances their powers and further encourages them to treat the rest of us instrumentally and manipulatively. But such confinement is precisely the method employed in modern societies to avoid the worst excesses of informal social control at the group level based on stigmatization and exclusion. Bureaucratic control and manipulation may sometimes be the lesser evil here. It may even be that there is too much anxiety about the rise of purely instrumental orientations to human beings, and too much criticism of new forms of knowledge and information that allegedly abet this but that have so far led to few if any clearly adverse consequences. Treatment as an object for purposes of bureaucratic processing is generally preferable to being treated as an object of hatred and animosity; and in many of our dealings with others it can be more important to be treated professionally than with the full respect due to a rational being.[5]

Even so, a long hard look at the future does reveal grounds for anxiety. Genomics and genomic technologies are still at an early stage of development, and their social deployment has scarcely started. But even now the gathering and storing of human genomes is proceeding at a frantic pace in all kinds of diverse and uncoordinated projects. Police forces are taking DNA samples from supposed offenders. Military bureaucracies have taken note of the utility of DNA in identifying corpses and acted accordingly. DNA is being collected by medical professionals in vast research projects intended in due course to improve health. In the UK, for example, the Biobank project is an enterprise designed to collect half a million DNA samples, from the middle-aged and elderly population, and to correlate the genomic information derived from these with medical records, including social and economic data as well as straightforward information on illness and abnormal physical conditions. Indeed, in the UK several million DNA samples are already in the possession of state bodies, and if current trends continue it will not be long before almost everyone in the country will be linked to this repository of information, through their relatives if not directly. And if we set alongside this growing collection of indefinitely storable DNA the ever-increasing rates of sequencing made possible by ongoing technological change, and the exponential growth both in administrative employment

5 Much more could be said here. Both sociologists and philosophers have expressed anxiety about the purely instrumental orientations to human beings characteristic of administrators and bureaucrats, but it can shorten the time spent at airports. And moving into the secret realm of highly skilled professionals, we can encounter highly functional respect-lowering devices: too much respect may make the hand of your eye surgeon shake; better perhaps that you are treated as an object for an hour or two.

and in the IT that amplifies administrative capacity, we are confronted with the onset of a sea change in the extent of the powers of the state over its citizens. What the consequences of such a major change in the magnitude and distribution of these powers will be is becoming a pressing question.

Ever increasing bureaucratization and regulation is a secular trend, long recognized and long explained as the inevitable consequence of the growing power and complexity of modern societies. It has been realized in the USA and Europe over many decades by a steady growth in the bureaucratic workforce of more than 3% annually, an increase that has continued unabated even when general employment has been in decline. The growth of our current information technology, widely agreed to follow Moore's law, a doubling in the capacities of most crucial components of the systems roughly every year and a half, further increases this growth of administrative power. And the growth of genomics both as knowledge and as technology in close association with this growth of IT is now yet another factor, facilitating the development of new and ever more sophisticated ways of sorting and processing people. And as if these developments were not enough, there is currently a new incentive to bring them together and mobilize them, the so-called War on Terror. The vast stock of genomic information initially gathered for other purposes now constitutes a standing temptation to those seeking political support by emphasizing the "terrorist threat." It is not inconceivable that DNA sequences marking dysfunctional bodily conditions may before long begin to attract attention because of their slightly increased frequency among inhabitants of the banks of the Tigris, or remote valleys in the Caucasus or Hindu Kush: indeed, this kind of thing could provide the political bandwagon that moves genomics to the very heart of a future "administered society."

Of course, the prospect of an "administered society" stunting and impoverishing the lives of its citizens looms large whatever happens in the realm of genomics. A cancerous bureaucracy representing the drawn-out mental suicide of whole collectives of fearful human beings, should it occur, will not be wholly explicable as the result of narrowly technical changes. Even so, the close relationship of the capacities being produced by genomics and associated sciences and the capacities currently desired by bureaucrats and administrators and their controllers deserves notice, as does the fact that, whatever its promise for the future, knowledge of human genomes is currently being used more to transform our relations with each other than to transform our bodily condition.

Resisting genomic powers

Let us now return to the powers of direct manipulation and control of genomes increasingly being provided by the biological sciences. It is not unrealistic to imagine an extension of these powers so great that we eventually become capable of reconstituting genomes much as we wish, of significantly modulating ontogenetic processes and associated cell chemistry, and of intervening in the detailed operation of the molecular systems associated with genomes in fully differentiated organisms. This is the situation toward which research is all the time moving; the prohibitions and regulatory restraints increasingly being placed on particular technical innovations, on cloning and stem cell research, for example, or the commercial growing of genetically modified (GM) crops, are likely to have only limited restraining effects, as indeed consideration of these very cases amply illustrates.

Insofar as other organisms are concerned, it will become possible not merely to modify their genomes to "improve" their phenotypes, but to treat them as biological factories within which to produce all kinds of needed materials—drugs, hormones, nutrients, vaccines, and so forth— or, alternatively, to modify them in ways designed to affect the environments they inhabit and the ecological balance of which they are part. Insofar as our own genomes are concerned, we shall be confronted with the much-discussed possibilities of remaking ourselves, not just via genomic manipulation but through tissue engineering as well, of redrawing the physical boundaries around ourselves, and of redefining both the social and biological relations that link us with each other.

If we look away from what some of us will hope are distant prospects and focus on the shorter term, several techniques are already emerging out of genomic science with enormous, though as yet largely unproven, potential to influence human health. Gene therapy aims to insert functional DNA sequences into genomes in which they are absent or nonfunctional.[6] As in more familiar processes used in the genetic modification of plants, viruses into which human genetic material has been introduced are beginning to be employed for genetic insertion, although a long series of technical difficulties and unexpected hazards has impeded their use in this context; the use of bacterial artificial chromosomes is also contemplated.

6 Gene therapy is generally defined narrowly, not as therapy on genes but as therapy using genes, and involving the insertion of new genetic material. On this definition PTC124 ingestion, as described in the next footnote, is not gene therapy, nor would the medical use of RNAis count as such.

At the same time a number of promising techniques are being developed to switch off or repair defective DNA sequences, including the use of chemical compounds that if taken regularly in tablet form promise to restore production of missing proteins and hence to control some genetic diseases.[7] So far all attempts at gene therapy have been directed towards somatic rather than germ cells, and the latter are of course much more controversial targets because success here implies the possibility of making permanent changes to the gene pool. On the other hand, correction of a genetic defect not only in a particular sufferer but in that person's descendants has obvious advantages. And since the technical problems are little different in the two cases, it seems inevitable that if and when reliable gene therapies are developed for the treatment of somatic cells, the question of germ-line genetic therapy will immediately move to the center of debate.

Currently, the most widely discussed therapeutic possibility deriving from developments in molecular biology is so-called stem cell therapy. As we noted in chapter 3, stem cells retain the potential to develop into all or most of the specific cell types that constitute a human body. The underlying idea is that such cells could be induced to develop into replacements for failing human tissues. So, for example, it is hoped that eventually stem cells could be used to grow new heart tissue to repair a damaged heart, or even new nerve cells to connect the severed parts of a broken spine. Procedures involving two main sources of stem cells are envisaged. The less controversial procedures involve so-called adult stem cells, undifferentiated cells that can be found in many human tissues. The most familiar and well-established example of a procedure involving their use is bone marrow transplantation for the treatment of blood diseases such as leukemia, but promising results have been claimed for a variety of other conditions, including cardiac disease. An interesting variant here is the contemplated use of stem cells taken from the umbilical cord at the birth of the individual needing treatment and stored until required. And here of course is yet another potential impetus to the growth of vast banks of stored individually identifiable genomic DNA.

Far more controversial are procedures involving the use of "embryonic" stem cells. These are cells derived from the early stages of development

---

7 One of the best known of these is PTC124, which appears to be effective in some cases of muscular dystrophy and other recessive disorders. PTC124 encourages the protein-synthesizing machinery to read through stop codons, and hence to continue indifferent to any point mutations that have produced spurious stop codons (Welch et al., 2007).

of the human embryo. The great attraction of these cells for research-ers is that near enough immortal lines of these cells can be established and can be induced to differentiate into practically any desired tissue type. Much of the controversy surrounding them has derived from the fact that establishing lines of embryonic stem cells has involved creating and then, in effect, destroying human embryos. Not everyone regards this as morally problematic, but it is an activity that evokes highly emo-tive ethical debates in several important cultural contexts. In the US the so-called pro-life movement is committed to the view that a human life with full human rights begins at conception, so that creating stem cells by destroying embryos counts as murder. In Germany, ethical sensitivi-ties stemming from the experience of the Nazi era have left a legacy of extreme suspicion of genetic/eugenic technologies and produced a legal structure that asserts an uncompromising respect for human dignity; and here too intense debate surrounds the possible medical use of em-bryonic stem cells.

Of course, a continuing broad advance in knowledge and power has traditionally been regarded as in itself a good thing. It represents prog-ress. It allows us to do many desirable things previously impossible but does not oblige us to do undesirable things. Above all perhaps, in the case of genomics, it promises improvements in health and life expectancy, and not just by correcting the text of our own genome. The genetic engi-neering of other organisms offers hope, even in the near future, of reduc-ing the incidence of major diseases like malaria. And given that deaths annually from malaria alone are currently measured in the millions, and that this is but one possibility being explored, there is evidently some-thing to be said for a belief in progress of this sort.

Even so, not everyone accepts this optimistic story. The new genomic technologies have attracted significant opposition just as all new tech-nologies do. And opposition to technological change is not merely an irrational opposition to progress. It can be backed with powerful argu-ments, even with powerful utilitarian arguments. The burgeoning pow-ers and capacities of advanced science and technology are subject to misuse and could potentially harm humanity as a whole. Sometimes the benefits of new powers and their application go to one group of people or set of institutions and the costs are paid by another, in which cases the "resistance to progress" of the latter can scarcely be accounted ir-rational. Similarly, the benefits of a technology may be enjoyed by one generation and the costs left to be paid by the next, as has happened, according to some, in countries that have deployed nuclear power with insufficient regard for decommissioning costs, and is almost certainly happening now as our profligate burning of fossil fuels leaves our descendants to cope with the legacy of global warming. There is also the important class of cases, particularly interesting in relation to genomics, where the individual benefits from her own exploitation of a power or technique, but incurs far greater costs from the exploitation of the technique by others in the collective as a whole. This notorious problem of collective action is nicely illustrated by recently developed powers to select the biological sex of a child.[8] In many societies the birth of a male child brings clear individual benefits, but if all the members of such a society were to select a male child, disaster would ensue for them, both individually and collectively. This simple example, of course, nicely captures a problem associated with very many eugenic techniques if made freely available to individuals: many of the imagined pleasures of a designer-baby would not be enjoyed in a society of designer-babies.[9]

The classical response to arguments of this sort is to claim that regulation and control may iron out these difficulties and ensure that powers are used for the collective good. If unfettered economic calculation does not itself lead to the optimal use of new technologies, then political power superimposed upon it will ensure that it does so. The trouble is, of course, that if technology is a double-edged resource, so too is regulation. We have belabored this point in the previous section. Blanket trust in regulation implies a Utopian conception of political power as a deus ex machina that has the well-being of humankind as a whole at heart, not to mention a questionable assumption that no unintended consequences will ensue from the use of that power. It is also important to remember that regulation and control carry costs of their own. Many would claim that the costs of regulating powers and preventing their abuse frequently exceed the benefits derived from developing and deploying them: nuclear weapons technologies are thought by some to constitute an extreme example of this sort. But in any case the notion of cost-free regulation and control is clearly nonsense, and there are vital

8 Perversely, but no doubt out of a well-meaning concern with user-engagement, the procedure is widely referred to by medical professionals as "gender selection."

9 We were initially unaware that this term, denoting a baby created to order with given specifications, connoting a fashion accessory, has established itself firmly only in the UK version of English. In any case, perhaps a better illustration of the same point is provided by the film *Gattaca*, in which a genetic class society is created, with the genetically "enhanced" increasingly divided from the genetically impoverished, but not to their benefit. See also Silver (1998).

issues here of what the costs are, how predictable they are, whom they fall upon, and how they are to be calculated.[10]

Just as there are costs associated with regulating technology, so there are also with implementing it and adjusting to it. And these last costs, which generally precede the benefits that the technology may provide, raise intriguing issues in societies like our own in which technological "progress" is continual and relentless. The fetishism that values progress above all else involves something very like identifying life on a building site as the highest form of human aspiration. Not everyone wishes to live on a building site. A common empirical finding is that once they have achieved a certain minimal standard of living, many people come to value the status quo above even apparently beneficial forms of change and may sometimes oppose advancing technologies as threats to the predictable and secure environment to which they have adjusted. Even in industrial organizations, and companies faced with the rigors of the marketplace, there is a strong inclination to "satisfice" rather than to "maximize," that is to continue routinely to reproduce a satisfactory status quo, rather than to act in a way that offers the prospect of greatest future profit. And similarly, for individuals in rich countries satisfied with their ways of living, having to adjust to change counts as a cost that offers a perfectly intelligible utilitarian rationale for opposition to scientific and technological advances. For them, such advances may represent a disruption that threatens to turn a loved environment into the equivalent of a building site to no good purpose. However, not everyone has reached the minimal level at which satisficing starts to be attractive, and many in poorer countries live as they do because they are unable to live otherwise and not because they have no wish to do so. And this may engender very different attitudes to technological change in rich and poor countries.

Within rich countries themselves, meanwhile, technological changes will affect different groups and even different individuals in very different ways, and the inclination to satisfice may vary between the more and the less affluent. Economists tell us that this need not matter insofar as the provision of divisible goods is concerned. The individual wants and needs of the more and the less affluent, of satisficers and maximizers, of

10 There are further costs involved in calculating the answers to these very questions, which indicates the existence of a major and unaccountably neglected feature of all cost-benefit analyses. They can never, ever be fully rigorous since the making of them adds a cost that they need to allow for, but in allowing for that cost another is created, and so ad infinitum.

early and late adopters of technology, and so forth, can all be separately met in a market system that permits them to be expressed as demand.[11] In the specific cases of genetic engineering and other genomic technologies, however, things are not always so simple. Many of the affluent consumers who fail to see how they themselves might benefit from new technologies are not disposed to wish them well elsewhere either. In the case of GM foodstuffs, for example, there is not merely a widespread disinclination to consume them but significant opposition to their development and cultivation per se. The attitudes of many opponents of GM differ significantly in this respect from other groups with specific dietary preferences such as traditional vegetarians, most of whom buy their vegetables in the market and leave others free to buy meat in the same place.

In Europe, including the UK, as opposition to the commercialization of GM foodstuffs has crystallized into something close to an organized social movement, so the rationalization of that opposition has extended beyond recognizably utilitarian concerns with specific costs and benefits into a general antipathy to all such products and the technology behind them.[12] The concerns explicitly expressed have, nonetheless, remained predominantly utilitarian. Certainly, this is true of Greenpeace's successful campaign against the growing of GM crops in the UK, begun in the mid-1990s, which stressed the risks of pollen release into the environment and, more generally, the risks of exploiting the new and incompletely understood recombinant DNA technology on a large scale. Greenpeace followed a standard line of argument here, employed by environmental movements in many different contexts, that stresses the risks associated with the deployment of new powers and the enormous costs that could ensue from ignoring those risks. Politicians and religious authorities have long relied on such strategies, invoking invisible risks and dangers as they sought to incite wars, pogroms, witch burnings, and campaigns against "threatening"—in fact often harmless

11 This is not true of indivisible goods, of course, where the collective action problem mentioned earlier arises.

12 In the USA, things have gone very differently and many of the arguments raised in the European context have failed to move audiences there. And were one to look to other parts of the world, a far more complex story would need to be told, reflecting cultural diversity, different degrees of economic development, and international political relations. Comparisons among societies are needed to enrich understanding here not just of difference and variation but of what is going on within the specific contexts being compared (Jasanoff, 2005).

and powerless—minorities. And economic organizations have employed similar rhetoric to sell their products as prophylactics, whether armaments or deodorants. But this same discourse is now the favored weapon of those groups and movements that oppose powerful institutions and resist the uncontrolled proliferation of the new technologies they rightly regard as associated with them.[13] It permits them to project their criticisms and reservations with impressive plausibility in utilitarian terms, as calls for an appropriately precautionary response to risk.

We ought to say just a little about the background to this development, which entails going back a quarter of a century or so and identifying a major shift in the nature of politics that was close to completion in many European countries by roughly that time (Inglehart, 1997). Whether as a consequence of the demise of the Soviet "threat" or otherwise, these countries entered a period during which their basic institutional arrangements ceased to be the targets of any serious challenges: politics based on representative democracy, economics based on capitalism and the institution of private property, morality of a meritocratic flavor, and relations between subcultures involving tolerance and peaceful coexistence came to be taken for granted as the appropriate frame in which life should be lived. Of course this secular pattern of change has not occurred everywhere, and even if attention is confined to the USA and Europe, it is clear that not all countries made this journey at the same rate and that major variations on the basic pattern exist even between different parts of individual countries. Even so, we believe it is fair to say that the UK, which we discuss specifically, was close to the end of this road in the 1990s and that important changes in the nature and location of political conflict followed from that.

Political radicalism, of course, did not disappear in these conditions, but it was obliged to reorganize and redefine itself, which it did in large part by a move toward single-issue politics. Instead of seeking directly to mobilize power for a radical reordering of society, it concentrated on the promotion of highly specific objectives, dramatizing criticisms of the status quo in a highly theatrical fashion, and securing widespread if weak public support for their remedy by skillful use of the media. The groups and movements involved in this kind of activity soon found that support was more easily enlisted by evoking fear and a sense of insecurity than

13 To recap what was said in the opening paragraphs of the chapter: the technologies, in the form of persons and their knowledge skills and artifacts, constitute powers, and the associated "powerful" institutions are powerful in that they have discretion in how those powers are exercised.

by appeal to moral and ethical principle, and the risks and dangers of technical innovations and projects became politicized as never before, to the extent that the technical realm was sometimes proclaimed as the new realm of politics itself, with the traditional political realm increasingly becoming an empty husk.[14]

One result of all this was that scientific expertise was set into a new and ambivalent relationship with some of its audiences, and different interests battled either to secure its support for their positions or, if they entertained no hope of doing so, to call the authority of that expertise into question. The credibility of science became a major political issue, debated in the wider public arena as never before. And controversy about the correct interpretation of scientific findings and their putative implications for major policies and political decisions also moved to center stage. Science was at once the basis of the knowledge from which the benefits of major projects were calculated and the source of the risk estimates relevant to the rational assessment of their dangers, which meant that it now had much to offer to both sides in any political/technical controversy. Radical and environmentalist critics of technological change now had a use for detailed scientific findings as much as did the established repositories of political and economic power to which scientists had traditionally proffered advice. And no scientist could legitimately refuse to acknowledge the existence of risks and uncertainties in any given case, since science can claim authority only as a source of empirical knowledge and with empirical knowledge there is no certainty.

Controversies in the UK about the commercial growing of GM food crops have been typical in these respects. They have focused specifically on the environmental and health risks of the crops, with both sides content to argue in a nominally utilitarian frame and happy to acknowledge the standing of science as a possible source of insight into just how serious the risks are. The two sides have differed, and continue to differ, in their accounts of the actual risks involved, of the technical evidence supporting their different estimates of these, and of what it would be sensible to do in the light of them. And the particular advantages conferred on the opponents of GM by their ability to frame the dispute as a controversy about risks and dangers have been apparent throughout. The very existence of a technical debate about the risks and dangers provided grounds for doubt about GM among nonexpert audiences: in the context of everyday understanding, where something may perhaps

14 For the invocation of risk by environmental and other oppositional movements Ulrich Beck's *Risk Society* (1988) is the locus classicus.

be risky, then ipso facto it is risky. And where the cited risks are risks with disastrous consequences, playing safe often makes sense, particularly where, as in the case of GM and UK consumers, it is perceived as an almost cost-free policy.

We want to say a little more about this controversy, but we do not propose to support either of the two sides or to adjudicate between their opposed arguments. Of course detailed analysis does raise questions about these arguments, and both sides have exposed weaknesses in those of the other. Advocates of GM foods have rightly been accused of downplaying the problems of risk and safety surrounding their products. And it has been claimed equally plausibly that these alleged risks and uncertainties have been talked up by activists intent on engendering the doubt that has been described as the characteristic product of the "social movement industry" and linking it to the direst of dangers specifically for the consumption of nonspecialist audiences.[15] More important to the present discussion, however, is how much the two sides in the UK debates agreed upon. Perhaps because the case for the safety of GM was largely developed in response to its critics here, there was an extraordinary degree of consensus on the agenda to be debated and on how the debate itself should be framed.

The public debates about the risks posed by the growing of GM crops were so prolonged and intense that it is easy to assume that they must have been comprehensive and wide ranging, but the narrowness of their agenda is apparent in how many themes and topics were neglected. In particular, relatively little attention was paid to how GM products differed from conventional products. And the processes whereby they were produced also received surprisingly little discussion. Far from being highly reliable and predictable, the methods used to introduce alien DNA into organisms are still very much hit and miss. Genomics has yet to be properly exploited in this context. What goes into seeds may be controllable, but the delivery systems are indiscriminate and unreliable, and where in the genomic DNA the new material ends up is often a matter of chance. Repeated stringent selection and extensive testing and quality control procedures are supposed subsequently to compensate for this, to stabilize desired traits in products and ensure the absence of unwanted ones. But neither the unpredictability of production processes where one pulls the trigger of the gene gun and hopes for the best nor the

15 See Zald and McCarthy (1987) for "the social movement industry" and for more general critical insight into the differences between movement activists, their passive supporters, and their wider audiences.

scientific rigor of subsequent selection and testing procedures featured prominently in the public debates.

Of course, even though it was largely constituted as a controversy about the risks of contact with a given category of materials,[16] whether in the form of pollen particles or food particles, the debate still covered more topics and raised more issues than we can possibly review here, and we want to focus our own discussion more narrowly still, upon just one key assumption that structured the exchanges between the two sides. Both treated the controversy as if it were essentially a technical one, about something that was really there, something that empirical research could aspire to identify—the actual risks surrounding the growing and consumption of GM food crops. Both proceeded on the assumption that perceptions of risk were linked to empirical evidence and technical argument and that further evidence and argument might change these perceptions and perhaps even resolve the issues in dispute by establishing what the risks really were. We need to question this assumption.

Arguments and institutions

How far the UK debates on the risks of GM are intelligible as technical ones wherein contributors based their opinions on evidence and arguments and revised them in response to more of the same is a more difficult question than it may appear. No doubt if large numbers of people consuming GM foods started dropping dead prematurely and disproportionately the foods would soon be marked as serious health risks, by many of us at least. Common sense rationalizes this with the thought that how risky GM really is can and does have a bearing on how risky people believe it to be. Even so, minds cannot be directly accessed and we can never be sure from speech or even behavior just what is moving them and affecting beliefs and convictions. An individual moved by one argument may invoke others as well, simply in order to increase support for her cause. A group or movement supporting a policy for whatever reasons may decide in private how to rationalize its position and which arguments to present to which external audiences in order

16 That the debate was constructed around unknown dangers at point of contact is interesting. The resulting discourse is thereby given a form analogous to that of familiar discourses of pollution, contagion, infection, and contamination. And indeed GM products are now increasingly being thought of in just these terms. More is said on this in chapter 7.

to attract allies. We cannot assume that its individual members believe these arguments to be valid, and are not merely recognizing the need to sing from the same hymn sheet. To put these points in an extreme form: in principle, none of the arguments advanced in public debates such as that concerning the risks associated with GM food crops need figure among those actually reckoned valid and relevant by the participants.[17] And hence any analysis of the nature of this and similar debates has to begin with the arguments themselves and to recognize that any conclusions about whether and how far they move minds must inevitably be speculative ones.

Let us then begin with the arguments. When anti-GM activists advocate avoidance of GM products, or even abandonment of the entire technology, as the proper precautionary approach to the risks they represent, they follow standard lines of argument now widely used, particularly by environmentalist movements, to attack a whole range of practices and policies.[18] But what is referred to as "risk" here is actually better described as "uncertainty." There is an important distinction to be made between the meanings of the two terms even though it is widely ignored and they have come to be used interchangeably in many contexts. Risks are generally understood as quantifiable; they are the predictable chances of things going the wrong way and generating a given harm. They are things that gamblers and insurance companies face with equanimity all the time, and profitably allow for, at least in the latter case. And they are faced similarly all the time in everyday life, as we get on the train or into the car, aware of the probability of an accident yet willing to chance one to make our journey. Uncertainties, in contrast, are unquantifiable possibilities of things going wrong and causing harm: we speak of them where we are aware that something *could* happen, but where we have no idea of what the chances are.

17 In the course of following the progress of the GM debates in Britain, we have been told on several occasions in behind-the-scenes conversations with participants on both sides that the safety issues ostensibly being debated are not what the controversy is "really about."

18 It is common here to find references being made to "the precautionary principle" as the proper basis for regulatory decisions. But there is no one precautionary principle. There are many, with different formulations being put forward by different groups according to their objectives. That some of these formulations are impossible to meet by any innovation at all, however carefully introduced and rigorously regulated, does not necessarily count as a disadvantage by those who favor observance of them.

There must always be uncertainties associated with the use of genomic technologies, if only because our knowledge of genomes must always be less than certain. We may use whatever knowledge we have in order to calculate risk, and take the risk into account in deciding how to act. But what of the uncertainty in the knowledge used to calculate the risk? There is no way of converting that to risk, save by using more knowledge, which brings in more uncertainty. Thus, while risks are calculable and may accordingly bear upon rationally calculated decisions, including decisions reflecting a cautious, precautionary orientation and not awaiting evidence of actual harm in a particular case, uncertainties are a different matter altogether. Meeting the escaped lion on the street, most people would take evasive action without waiting for it to take a bite out of them. They would not await evidence of the lion's unfriendliness before taking precautions but would act on the basis of knowledge of lions generally. And they very reasonably expect regulation to be similarly ahead of the game when there is reason to believe that significant risk exists in a specific case. But it is not clear what counts as a rational precautionary response to uncertainty, as opposed to risk.

There is no greater nightmare for regulators, and the political systems that they answer to, than that of finding acceptable precautionary approaches in instances surrounded by controversy, where one or other side is highlighting uncertainty. There are no arguments here with which regulators can dispose of their responsibilities, and the fact that the contending parties putting pressure on them may have no good arguments either does not necessarily compensate for this. How do they ensure the safe operation of nuclear reactors? The knowledge of theoretical particle physics may have to be drawn on here, but it is not a settled and complete body of knowledge: physicists disagree on fundamental issues like the invariance of the speed of light, and the Higgs boson remains worryingly elusive as we write. What is a sensible precautionary approach until physics sorts itself out? The trouble is, of course, that we can calculate rational policies only on the basis of knowledge, and this renders us impotent in the face of the uncertainty intrinsic to that knowledge itself. But of course things are no better where knowledge is lacking altogether. What was a sensible precautionary approach to the outbreak of bovine spongiform encephalopathy (BSE, or mad cow disease) in cattle in the UK? It is only a slight distortion to say that there was initially *no* knowledge relevant to assessing the chances of the disease jumping species and affecting humans. The agent itself was unknown. The empirical analogies with familiar cases were slight. Although scrapie, an apparently similar disease of sheep, gave some cause for comfort in that it was unknown in humans, overall there was scarcely anything to go on. Eventually some billions of pounds were spent (money that might have saved thousands of lives if otherwise deployed) responding to a hypothetical possibility, with no idea at that time of how likely the possibility was, or how consequential, or whether the expenditure would significantly alter the chances of its occurring.

In most controversies surrounding major policy decisions, one or both sides will be found invoking a cocktail of risks and uncertainties, usually all described indifferently as risks. The actual balance of risk and uncertainty tends to reflect existing knowledge of the situation, or as it is commonly put, the available evidence relevant to it. Thus, in the controversy about how far human agency is currently responsible for climate change, arguably one of the great technical issues of our times, both sides cite evidence and even agree on the good standing of some if not all of it, so that risk, as opposed to uncertainty, has a prominent place in the debate. And when in 2002–3 the UK government controversially decided to participate in an attack upon Iraq, on the specific pretext that weapons of mass destruction (WMD) existed in the country, evidence for and against this claim was also cited by both sides, although here there was much less common ground. Still, how the two sides treated "the evidence" may have helped audiences to decide which side to lean toward. The peremptory dismissal of the evidence of UN weapons inspectors suggesting that the alleged WMD were not actually there was a significant episode in the controversy in the UK that adversely affected both the credibility of the government case for invasion and its overall credibility subsequently. But in contrast to both these cases, the outcome of the UK controversy over GM crops was effectively decided without major disputes over evidence and hence without major confrontations over risk in the narrow sense.[19]

Conceivably, this had something to do with the many millions of tons of GM foods that had been grown and consumed already in the US and elsewhere without obvious harm either to health or environment. But in any event the UK debate over the "risks" of GM crops focused heavily on uncertainties rather than risks in a narrow sense, with one

19 In 2003–5, arguably too late to affect outcomes, the results of lengthy comparative crop trials finally appeared and were used to justify political decisions that had the effect of killing off, at least for a time, the hopes of the pro-GM lobby. However, how far these results should have been taken seriously as evidence with any relevance at all to the general controversy over GM is itself controversial.

side depicting them as specific to GM and threatening to be costly and the other questioning the grounds of both claims. Thus, one side continually spoke of the unknown and uncertain effects of genetic engineering; but the other asked why these as yet unknown effects should raise any greater anxiety than the as yet unknown effects of recombination by the sex mechanism and standard plant breeding techniques. One side pointed to the danger of engineered DNA escaping into the environment; but the other noted that we had never worried in the past about the risks of naturally occurring recombinant DNA diffusing from place to place and escaping from human control.[20] One side associated the unquantified "risks" with costs so extremely high that they would outweigh any possible benefits whatever the odds of their having to be paid actually were. Perhaps they had recalled Pascal's famous advice to bet on the existence of God, because the loss of the infinite benefits of heaven, even if the chances were a billion to one against, still made it irrational to bet the other way; but the other side, possibly less well read in philosophy, did not buy the argument. One side argued that since they were "new" products greater risks and uncertainties were attached to GM crops than to standard ones; but many on the other side refused to accept that these products were new in any relevant sense or that the risks and uncertainties associated with them were different in kind from those associated with "natural" materials.[21]

Let us pause at this point and recall why we are interested in these arguments and the debate they belong within. Our question was whether the debate could be regarded as for the most part a technical one about what empirically was the case. The mere fact that participants tended to stand on one side or another throughout, contributing to what it is reasonable to call pro and anti positions, could be taken to imply a "no" answer. And further support for this conclusion can be derived from

20 Some on the pro-GM side also recalled in passing that when genetic engineers had developed techniques that more or less eliminated the risk of escapes of pollen into the environment, they had been denounced for restricting the use of their products with "terminator genes."

21 The uncompromisingly naturalistic view that admits of no fundamental distinction between GM and ordinary "natural" products is often taken to justify the principle of substantial equivalence that currently guides their regulation in the USA. This principle offers, on the face of it, an attractive solution to the problems associated with risk and uncertainty. Let us be even-handed in how we cope with them and scrutinize all products alike for possible harms, using the same standards and procedures. Stringent regulation may indeed then be indicated for some GM products but not merely because they are GM.

the fact that polarization persists today and that many of the same campaigning groups can be found still making the same notably different assessments of risks and uncertainties and never being at a loss for an argument with which to respond to the criticisms of those on the other side. But we have also sought to open another route to much the same conclusion by reviewing the opposed arguments themselves and noting how often they involve different judgments of uncertainty, as opposed to risk. Even if these different judgments are indeed authentic expressions of the convictions of the involved parties, it is difficult to see how arguments about which are the more credible can be understood as technical arguments about what is the case empirically.

Since most of us are in the habit of looking to science for authoritative guidance on empirical matters, we may well think it reasonable to look to science to estimate the risks associated with GM, even if it remains up to us or our representatives to determine what account we shall take of them at the level of policy. In the case of uncertainties, however, science can tell us no more than we know already: it can agree with us that they exist but can tell us nothing either of their magnitude or how best to take account of them. Uncertainty is indeed a peculiar notion and extraordinarily difficult to handle. We cannot do without it; if we know anything at all, it is that we are fallible, and that complete certainty about any empirical state of affairs is something we should never claim. But we cannot do anything with it either, or rather we cannot rationally do anything with it. An uncompromisingly precautionary approach to uncertainty—never act until you are certain no harm will be done—would entail total paralysis. On the other hand, a selectively precautionary approach applied only to particular areas of uncertainty has no defense against the criticism that it is arbitrary and inconsistent. Even so, we often feel obliged to act in the face of specific uncertainties, even in the absence of any clear justification for any given response to them, and uncertainties are often invoked, notably by opponents of technological innovation, precisely in the hope of provoking a particular kind of response.

As we argued earlier, the invocation of risk and uncertainty has proved an invaluable rhetorical strategy for those opposed to innovative technologies, and the recent campaign against the cultivation of GM crops in the UK is one illustration of its successful use.[22] Appeals to uncertainty have enduring attractions as bases for arguments that

22 Although the critics won that battle they may well have lost the war, however. Their favored forms of argument were evidently not sufficiently compelling in the USA.

can never be refuted. Let a disaster occur and those who warned of it are vindicated and revealed as prescient. Let all go well and uncertainty remains notwithstanding. No amount of evidence or counterargument will suffice to defeat the case that stands on uncertainty: if it is indeed uncertainties, and not risks, that are at issue, then evidence cannot bear on how we weigh them. But what then does bear on the weighing of uncertainties; for it is clear that some uncertainties are in fact given greater weight than others? Indeed it is clear empirically that people single out uncertainty for special attention in some contexts, while remaining oblivious of it in others, and that systematically different accounts exist of where significant uncertainties lie, and which associated threats and dangers societies need most urgently to address.[23] As to why this is so, one plausible conjecture is that uncertainty is invoked out of expediency. Uncertainty elicits fear and anxiety. Accordingly, while advocates of a new technology may prefer to make no mention of uncertainties, their enemies will associate them with the technology in order to put the frighteners on and encourage audiences to "play safe" and shun it. If this is the case, however, a further question needs to be answered. Why do the frighteners frighten? Or, more precisely, why do they frighten those who are frightened and fail to frighten those who are not?

Utilitarian arguments surely affect people and may help to bring about changes in what they believe. But here is one kind of utilitarian argument at least that confounds our standard intuitions of how arguments affect people. The traditional image of inferences from evidence moving our reason in the right direction will not serve here; references to uncertainty must move people differently, if they move them at all. Perhaps those affected by them are affected, not through any detailed calculations they are led to make, but in a less specific way, through other causal pathways; and perhaps this is why different groups of people are affected differently. What this suggests in the specific case of the GM debates in the UK is that the debates should not be addressed on the assumption that they were a series of narrowly technical arguments, but treated in a broader framework as contingent historical events.

One influential approach of this sort starts by noting that all of us operate on the basis of our received knowledge of the social order around us. We use an internalized cognitive map, let us say, of the institutions

---

23 People also respond very differently to different quantifiable risks. They are typically far more sanguine about the risks involved in driving their cars than about the generally much lower risks of traveling by plane or being victims of terrorist attacks, where their fate is less in their own hands.

and organizations we have to relate to, typically marked with prior indications of which may be relied upon and will repay commitment, and which need to be treated with suspicion. These received maps will vary from one institutional location to another, possibly as a result of experiences accumulating over long periods of time. It may be, for example, that those in the dominant institutions come to see the margins of society as the most likely sources of danger, whereas those on the margin especially fear the center and its daunting powers.[24] Thus, it may be that those in both locations are largely immune to some revelations of uncertainty and hypersensitive to others, and that their different sensitivities have a great deal to do with their different opinions on the safety of GM foods. If this is so we can no longer regard these opinions simply as different technical evaluations. We must accept that they are likely to be inspired as much by prior evaluations of the institutional carriers and controllers of the GM technology and its products as by the technology itself. Who is messing with genomes will now be at least as important as the fact that genomes are being messed with. The social distribution of the new powers of genomics will be liable to affect how they are evaluated as much as the actual characteristics of the powers themselves.

There is much to be said for analyzing the controversies surrounding GM food using this kind of framework, which can be further elaborated by noting how people in different institutional locations tend not only to operate with different received maps of the social order but with different received maps of the natural order as well.[25] These maps are liable to differ from those of specialists. It may be, for example, that for most specialists no fundamental distinction exists between naturally recombined DNA and engineered DNA so that inferences can routinely pass from the one to the other, both being instances of "the same stuff." But everyday knowledge sometimes marks out a strong distinction between "natural" substances and "artificial" substances or "chemicals." And similarly, insofar as everyday understandings have incorporated genetic notions, they may tend to genetic exceptionalism and not know of genes what chemists and molecular biologists assume is known of them. Consequently, in the context of everyday understanding, inferences to "non-

24 The work of Mary Douglas (Douglas, 1992; Douglas and Wildavsky, 1982) is the fount and origin of this kind of analysis.

25 Following Sheila Jasanoff (2004, 2005), these different kinds of map can be regarded not as independent of each other but as coproduced representations that are engendered, preserved, and transmitted together in the context of everyday life by the same social processes.

natural" from "natural" DNA may not be valid, and the uncertainties associated with the one may be reckoned more of a menace than those associated with the other, undermining the credibility of an important part of the pro-GM case.

Of course, scientists command high credibility, and people are generally content to set the opinions of scientific specialists above their own everyday intuitions. Ironically, however, scientists themselves may have encouraged an exceptionalist way of addressing the issues that has lowered the credibility of their own expertise in this case. For scientists made a crucial contribution to the creation of the category of "GMO," which was designed for use as a regulatory category more than as a means of identifying a particular kind of organism, but was easily understood in the latter sense and reified as a kind of entity possessing exceptional intrinsic characteristics. Moreover, these new kinds of thing were defined in such a way, by reference to the esoteric recombinant DNA technology used in their creation, that GMOs were identifiable as almost invariably the possessions of a few very large and powerful international companies. Thus, a near perfect target for campaigning organizations was created, specific to the Monsantos of modern global capitalism, companies that the campaigners delight in attacking and which are marked in red on the institutional maps of many of their sources of support.

7

Natural Order and Human Dignity

The order of things

Genetically modified (GM) organisms, and particularly
GM food crops, featured prominently at the end of the previous chapter, where their supposed risks were discussed
in a utilitarian framework, so let us continue to take them
as examples and look now at nonutilitarian objections
to their creation and use. It is interesting to recall here
that, although the successful attack on the commercialization of GM food by Greenpeace was mainly presented
in utilitarian terms, treating risks as costs, it never cited
any clear evidence of costly environmental consequences,
or adverse health effects. It did, however, increasingly refer to "Frankenstein foods" and this seems to have struck
a chord. Genetic engineering is frequently attacked as an
affront to natural order, and, where it targets the human
genome, as a threat to human dignity. These kinds of objection may well capture the basis of some of the intuitive
aversion to GM that has contributed to its rejection in
Europe.

Many scientists and medical researchers are deeply
suspicious of nonutilitarian arguments, which they often
experience as expressions of emotional revulsion directed
against beneficial technical advances. Privately, they talk of
a "yuk factor" and of irrationality that has to be overcome.

But while those who resist such arguments rightly emphasize that they are not logically compelling, neither are utilitarian arguments logically compelling. The present discussion will merely continue to explore how noncompelling arguments are actually deployed in policy debates involving technical issues and why reputable arguments of both sorts may be found powerful and persuasive by some audiences, while others remain unmoved by them, or even unable to understand why they should be moved by them.

Let us leave these issues for later, however, and look first at arguments that oppose the creation of GMOs because they threaten the order of nature. Intuitive responses of aversion to GMOs seem frequently to have expressed a sense that they are unnatural and do constitute such a threat. And these intuitions have been rationalized by accounts that identify GMOs as anomalous when set against the backdrop of the order of things as it is presently defined. Given our existing taxonomy of species, the GM organism as a whole is identified as an intermediate form that does not belong. Given our existing conceptions of the structure and function of an organism, in which every part makes a necessary contribution to the harmonious operation of the whole, the "alien" DNA of the modified organism is similarly anomalous. A transgenic organism may accordingly be perceived as a threat to both of these ways in which biological order is currently understood. And although arguments based on these perceptions have not been as prominent in the media and ordinary public discourse as have those based on supposed threats to health and environment, they have nonetheless been systematically developed, particularly by philosophers.

Taxonomic anomalies may be sources of all kinds of practical inconvenience, and may engender annoyance on that basis alone, but that is a matter of *utility* and not relevant here. The question here is whether the anomalous taxonomic status of something might of itself move people to call for a prohibition to be set on it. And on the face of it the answer is that it might. There do seem to be many people for whom GMOs are not merely anomalous, but also offensive to their aesthetic sensibilities because they are anomalous, though this way of putting the matter stands in danger of failing to convey how intense such responses to anomaly can be.[1] Scientists who have to deal with the yuk factor are forced into an awareness of this, although it has been anthropologists

---

1 Immanuel Kant famously held that the beautiful was the sign of the morally good. Nowadays it is more common to think of aesthetic judgments as personal and idiosyncratic, and of no moral relevance.

who have documented it most extensively and reflected on it most pro-
ductively (Douglas, 1966, 1975). "Yuk" is the routine, immediate, un-
rationalized response to dirt, but, as anthropologists stress, dirt is not a
particular sort of matter, it is matter out of place, matter that pollutes,
matter that purely by virtue of where it is encountered signifies disorder.
The worm in the soup is yuk, not the worm in the garden, and for some
of us the same is true of the mud on the carpet; body parts that would
elicit no attention at their normal points of attachment may induce re-
pulsion and anxiety if encountered elsewhere; even health-giving im-
plantations of blood or bodily organs may engender analogous disquiet,
particularly if species transfer is involved. Engagement with just a few
of the countless examples of this sort suffices to confirm the remarkably
intense reactions that encounters with disorder are capable of eliciting,
and to make it credible that disorder at the level of genomes might have
a similar effect.

Insofar as the production of GMOs is denounced as a transgression
of the natural order, we need to recognize that the claim may be authentic, that those making it may be rational human beings, and that
what moves them to such a strong condemnation may be a comparably
strong felt experience. None of this, however, justifies the claim itself,
which still needs evaluating in the usual kind of way. What is the order
that GMOs transgress? It must presumably be an order established in
genomes, and more specifically in their four-letter nucleotide sequences.
But these sequences vary profoundly and significantly from one individual to another, and continually change even within individuals as
both biological processes and human activities affect them. They are
changed as we cook, eat, and ingest them; as we recombine their DNA
in the course of sexual activity; and as we expose them to viruses, environmental chemicals, and natural radiation. And the natural order they
are assumed to represent is in any case often identified as defective and
pathological, for example by tests for genetic diseases. Why then is *this*
disorder not seen as deeply troubling? How are we to understand the
highly selective negative responses to disorder in evidence here?

Let us raise the same question in another way. Our sense of order
and harmony in nature derives from the specific features of it we have
chosen to highlight and to build our taxonomies upon, and many different choices are possible here. Accordingly, our sense of disorder and
anomaly derives not from nature itself directly but from the highly specific sense of natural order expressed in our use of given taxonomies and
schemas. So to grasp why GMOs are significant anomalies we need to
reflect on how we currently order and divide the species of living things,

and also on how we understand the internal order of individual organisms of whatever species. But our sense of the order of species is currently defined by a "Darwinian," "evolutionary" scheme, largely a taxonomy of distinct species, but still a taxonomy recognized as the outcome of changes, encompassing objects that are still changing, objects indeed in which the very kinds of change that brought them into existence are ongoing (Hull, 1992; Dupré, 1993). And our sense of the internal order of an organism, for all that it is based on the image of a stable functioning system, also understands that system as a life cycle involving conception, development, reproduction, and death—manifest changes that also entail systematic reorderings at the level of the genome. In brief, the particular taxonomies and schemas currently deployed in biology are explicitly recognized as devices for capturing transient patterns in a nature that in the last analysis is constituted of flux and process. So why should some of us regard the particular changes introduced into this flux and process by transgenic modifications as anomalous?

It has often been noted how easily we forget the changing context in which biological species taxonomies are applied and imagine that they are grounded in unchanging essences. And this suggests that when genomic technologies are assailed as disruptive of order and harmony, the order and harmony in question must be sought neither in nature itself nor, if these are properly understood, in the conceptual schemes we have constructed to make sense of nature. It will be an order and harmony that reside only in quite specific understandings of those schemes, in fact in reified and essentialist interpretations of them that are hard to reconcile with their actual role in biology. To put the point a little more forcefully, the "natural order" that genomic technologies threaten is the continuous creation of its critics.

Those of us who are distinctively sensitive to disturbances of the natural order are also likely to have distinctive conceptions of what that order consists in. And this means that debate about the salience and seriousness of threats to natural order will normally also extend to the issue of what precisely is being threatened. But even now it may be that we have yet to arrive at the heart of the matter insofar as responses to GMOs are concerned. What is perhaps most striking about these responses is how in the face of so many possible sources of disorder, it is *people acting upon nature* that are almost invariably the foci of anxiety. Of course even here the responses remain highly selective, and it needs to be asked why some of these actions are focused upon but not others. Why, for example, does the reordering of DNA via sexual activity go largely unremarked and the systematic reordering of genomic DNA by plant and animal breeders, frequently involving transfers between species, attract only scant criticism, when the reorderings effected by "genetic engineers" are so often found deeply repellent?

How far the human mind is drawn to pattern, order, and harmony, and naturally inclined to assign an intrinsic value to it, is an open question that we need not go into here. Even if there is some kind of universal aesthetic involved in our responses to disorder, in practice such responses, or at least those that figure large in the public sphere, are very strongly socially structured, and the rationalizing arguments they inspire and underpin are even more so. We need not deny that patterns may be attributed to genomes and DNA sequences, and that these offer many opportunities to those who valorize order and harmony and deplore its subversion by anomaly and pollution. But we need to acknowledge, nonetheless, that the most powerful collective responses to disorder are always more than responses to disorder per se.

We have just discussed recent and effective campaigns by anti-GM activists against "Frankenstein foods." The image of Frankenstein repels both through its confusion of human artifice and natural order and because it is a whole consisting in parts that do not naturally belong together. It invites us to think of GMOs as analogous to monsters and mythical beasts, and to transfer all the traditional responses of aversion to these anomalous creatures over to the genomic level. But it is intriguing that the analogy has to be made between transgenic plants and anomalous animals, whether human or nonhuman. Reference to existing transgenic plants might lead us all too easily into the order and harmony of the rose garden, and make us aware of how beauty and aesthetic delight may emerge from species crossing and hybridization. It is not easy to be repelled by the monstrosity of tea roses, although it needs to be remembered that aesthetic experience remains socially structured even in the rose garden. We, the authors, happen to share the preference typical of gardeners in certain social groups for species roses, and are happy enough to agree that many hybrid forms are the products of aristocratic excess or else of bourgeois ostentation. But we strongly doubt whether this constitutes an intuitive aversion to the disruption of natural order.

Long ago now, Sherrington (1940) famously reflected on the exquisite orderliness and beautiful interplay between structure and function revealed by study of the malaria parasite and the way that it was adapted to its particular niche in the larger scheme of things. His concern was actually with the argument from design, still taken seriously even then as a justification for a theistic worldview. But his account makes clear, nevertheless, how someone moved purely by a love of order, integrity, and harmony might conceivably derive enormous aesthetic satisfaction from contemplation of this remarkable natural object. No doubt there is a similar order and integrity to be contemplated in the genome of the malaria parasite, as also in that of the mosquito that is its favored mode of transport. And no doubt the way that the parasite inflicts debilitation and death on the millions of its human hosts, but with due moderation as a good parasite ought, and even the way that the host has evolved various not entirely cost-free defenses against the predations of the parasite, can also be contemplated as wonderful manifestations of the order and harmony of nature. But in practice these manifestations have encouraged attacks upon the natural order with the spray gun, and latterly with the new genomic technologies. In due course genomic monsters may make life hard for the beautifully adapted creatures that presently kill over a million of us annually. And indeed it is likely that Frankenstein flies and perverted Plasmodia will eventually secure widespread acceptance and toleration, even in those societies marked by their recent reluctance to tolerate GM foods.

Our responses to disorder and anomaly, then, are strongly socially structured. Generally, they are elicited by threats to our dominant systems of classification and the generally accepted ways of applying them, and structured in this way they are protective of the existing institutional order. But of course these responses may also be structured in other ways and be aligned against the institutional and ontological status quo. They may, for example, be elicited by social movements and pressure groups and ordered and intensified by the agendas these groups and movements promote. Responses to the anomalous status of GMOs have been elicited in this way, and ordered around a vision of scientists as originators of threats to the natural order: a simple variant on a familiar template, already widely disseminated, in which external human agency acts upon a benign nature and threatens it with chaos and disruption. Dr. Frankenstein is, of course, the canonical exemplar of such an agent.

Another way in which our perceptions and responses may be directed and structured derives from our vivid awareness of, and protective attitude towards, our own bodies and their boundaries. Indeed, the embodied human being as a constitutive part of the natural order and a potential victim of the external powers and forces that threaten that order is probably an even more important organizing template than that of the human being as external power acting upon nature, although of course the two are not mutually exclusive. The genetic engineers who, suitably assisted by fireflies or jellyfish, produce luminous plants that we may harvest in the hours of darkness may elicit revulsion in some and an avoidance of their products. But the thought that the production of humans who glow in the dark is also perfectly feasible technically is likely to be still more disturbing and repellent, even in societies like ours that are currently under serious pressure from environmentalists to reduce the energy consumed for lighting.

At the start of this section we mentioned Greenpeace's campaign against GM food crops and hinted that the case argued may not have been what actually secured the success of the campaign. In brief, whereas Greenpeace stressed the alleged environmental risks and uncertainties surrounding the crops, and the possibility of gene-transfer from them, one outcome, and perhaps the decisive one, was that many people became anxious about the safety of the crops as foods and refused to purchase them. No doubt a part of this response was utilitarian and precautionary: there seems to be some fuss about GM foodstuffs; best not to eat them until we learn more about what is going on. But it is interesting how, once GM was identified as alien and unnatural, refusal to allow it to enter the human body became crucial, not refusal to allow it to enter the environment, even though the latter was what the campaigners overtly sought to elicit with their utilitarian arguments. In our society, if not in all, the experience of threats to order—of pollution, adulteration, contamination, defilement, loss of purity and integrity—is most intense at the boundary between the human body and its environment. In what we eat, we give particularly strong expression to our sense of natural, moral, and institutional order. Our most powerful way of communicating disgust is to vomit (Douglas, 1966).

Ironically, perhaps, all this is currently a source of consolation, even of hope, for some of those on the losing side in the European GM debates. For all the reaction against GM and GMOs generally, and for all the utilitarian rationalizations offered in support of it, in practice a refusal to ingest GM food crops has been its only widespread public expression. Retailers do not proclaim their products free of GM cotton. And neither has the use of microbial GMOs in the production of drugs, vaccines, and other precious complex molecules attracted significant opposition, possibly because it is not the GMOs themselves that are imbibed or injected. Not even the insulin produced by GM *E. coli* bacteria—*E. coli*–human hybrids they could perhaps be called—has proved difficult to market, although some minor technical questioning of whether the new insulin is quite as effective clinically as the old has occurred. Thus, it is predicted that the genomes of both plants and microbes will increasingly be exploited as natural factories in which all kinds of valuable complex molecules will be produced cheaply and in quantity.[2] And having become familiar and gained acceptance and trust in this way, there is then a chance, or so it is thought, that they will go on to gain acceptance as foods.

Dignity

Aversion to anomaly and taxonomic disorder is rarely a purely abstract response, because classifications and actions are closely coupled. Only where we can make clear distinctions between things do we have the basis for confident coordinated action in relation to them. In particular, the differences among kinds of things highlighted by a taxonomy typically inform and legitimate different ways of treating them. But it is not merely the kind of thing an object is that affects how we treat it; it may also be a matter of where the kind is placed on a hierarchy or ranking. Hierarchical taxonomies of the kinds of living things were long understood as rankings of worth or status, with different ranks meriting more or less respectful treatment. "Higher" creatures appeared at the top of the page, close to *Homo sapiens* at the apex; "lower" creatures were placed further down, with the lowest little removed from the inert matter at the foot of the page. In between, intermediate kinds and categories were fashioned into great chains of being that might once have run from God and his angels, to higher and lower humans of many kinds, both "natural" and "social," and thence to animals, plants, other living things, and ending with nonliving objects.

Chains of being are diverse and complex constructions, but all are underpinned by the same key contrast, that between the material and the human, with intermediate ranks taking their standing from their location between that which is base, profane, and of no particular account, and that which is elevated, sacred, and to be treated with special respect. Wherever such hierarchies are recognized, aversion to the disorder and anomaly produced by the mere mingling of kinds may be intensified by a sense that what is elevated is being defiled and debased by its confusion, whether in thought or action, with what lies below it. Indeed, it could conceivably be that without hierarchy there would be no aversion to disorder—that aversion is not merely intensified by hierarchy but experienced at all only where mingling and boundary transgression

2 See also the discussion of synthetic biology in chapter 3.

occur between things set on different levels.[3] Either way, however, the current aversion to the manipulation of human genomes, and efforts to brand them threats to human dignity, can be understood a little better against this background.

We need to ask at this point just what is meant by "dignity." It is striking that while a considerable range and variety of procedures are confidently identified as threats to human dignity—cloning, the creation of designer babies, gene therapy, and many others—the notion itself is not often elaborated upon or clarified. It is one of those notions that we all seem to understand yet few of us are able to put into (other) words. The dictionary tells us that "dignity" denotes "elevation," and that to dignify is to invest with high rank or honor or elevation. This leaves many issues unresolved, but is useful in that it confirms that dignity is neither an observable property of things nor something inherent in them but rather something conferred on them by human beings: it is a status rather than a natural characteristic.[4]

The dictionary also leaves little doubt that in the paradigm case dignity is conferred upon human beings, indeed upon fully developed human beings, even if the actual use of the term is more diverse and complicated. There are of course other things besides human beings to which the term is applied, and other stages of the human life cycle to which its application may extend. Indeed, debates about how widely to extend the term within the human realm long predate current controversies about whether the embryo, the zygote, and the genomes the zygote contains possess human dignity. Longstanding debates about the respect due to the bodies of the dead remain salient in controversies surrounding organ transplantation, and enduring conflicts over the standing of living but as yet unborn humans continue in current debates about abortion. There have also been innumerable controversies about the different levels and degrees of dignity due to different categories of viable human beings, but in societies like our own these have largely disappeared and today the

3 The conjecture is a fascinating one but difficult to check. It does offer a possible explanation of attitudes to tea roses, mongrel dogs, and mules, and not only of the normal lack of aversion to these admixtures but also of the very occasional responses of aversion that mutts, for example, do elicit.

4 Recognition of this is rarely a problem today, but to our eyes there is a confusion of these things in medieval chains of being, where some of the links might be natural kinds of animals and plants, as we would say, and others the different rankings of people in the social hierarchy.

idea that individual persons should be accorded anything but an equal dignity is widely regarded as a deeply offensive one.[5]

Interesting though these complications are, however, our present purposes are best served by returning to the paradigm case. Dignity is conferred, paradigmatically, on human beings by other human beings. But what is conferred when dignity is conferred? Fundamentally, a status is conferred that identifies those possessing it as entitled to a special kind of treatment and sets them apart from ordinary, profane, physical objects. To treat human beings wholly as instrumentalities, as things that can legitimately be manipulated to serve the ends of others, as things lacking in any ends or intrinsic value of their own as is the case with material objects, is to deny them dignity.[6] Where human dignity is acknowledged, and as we shall shortly argue this is almost everywhere, this introduces restrictions alien to the purely instrumental orientation we have to (other) material objects.

Assignments of dignity restrict the scope of our ordinary instrumental means/end frame of reference, but at the same time they normally activate a second frame, one that structures the interactions in which we engage in the course of our social life. This frame is explicitly reflected upon much less than the instrumental one, but it is no less familiar to us through direct practical engagement. In this frame we treat others not as objects open to exploitation but as participants with a standing on a par with our own, in a system of interaction wherein we exchange knowledge and information, evaluations and judgments, ideas and understandings, without any prior calculation of the costs and benefits. Indeed, we do more than merely exchange these things; we keep them constituted as collective possessions, understood alike by everyone through shared

5 This is so despite our constantly conferring different degrees of respect on each other in the course of social interaction and recognizing gain or loss of dignity intuitively at that level. There is one sense in which human dignity is inalienable and another in which it can be lost by slipping on a banana skin. This is partly intelligible in terms of history: "dignity" long implied degree of elevation, and the current notion that every human being is possessed of an equal dignity is a variant that came to the fore in Europe only with the advent of the enlightened absolutisms of the eighteenth century.

6 Action in accord with the fundamental (or "categorical") imperative to treat humans solely as ends was of course famously enjoined by Immanuel Kant, and the less individualistic variant valorized by Jürgen Habermas in his account of communicative action is now scarcely less well known.

use, and hence available for further use by all in the relevant community. In this way, what we do together in an interactive, noninstrumental frame constitutes and sustains much of what must be accepted if we are to make calculations, and act collectively on the basis of their outcomes, in an instrumental frame. Collective instrumental actions depend on the interactions that engender and stabilize shared concepts and shared knowledge (Barnes, 1995, 2000). The interactive frame, wherein there is mutual recognition of human dignity, is thus far more than a secondary frame, a mere restriction on the scope of instrumental action. It is better regarded as the primary frame in the context of which our social life is conducted and the cultural resources it requires are fashioned and maintained. It is the instrumental frame, in which our actions in relation to profane objects are informed by calculations that employ those cultural resources, that is secondary.

Activity in an instrumental frame crucially depends on interaction in a different frame, a frame wherein participants treat each other with respect. And in this lie the beginnings of an understanding of why the acknowledgment of human dignity is a necessity and why we naturally look to it as the paradigm of dignity: only those who confer the requisite dignity on each other can operate as competent members of a society, whether interactively or instrumentally. Human beings are social creatures, always and everywhere encountered as members of some society or group; and their disposition to confer dignity and accord respect is what their sociality fundamentally consists in.[7] Nonetheless, today we tend to speak of dignity almost entirely in an individualistic idiom. Where our concern is to highlight the need of the individual for respect if she is to become a full participant in society and to make the most of herself as a member, this does not constitute a problem. But we are liable sometimes to be blinded by our individualism to the extent that we forget about the collective dimension of dignity and respect altogether and even fail to remember that in order to be possessed, dignity must be conferred. Many philosophical accounts of dignity come dangerously close to forgetting this. They characterize human beings abstractly as independent individuals. And they reify the dignity that interacting members of collectives confer on each other into

7 It goes without saying that this does not imply that always and everywhere humans treat each other with enough respect in an ethical sense. We speak in a functional sense of what is necessary to constitute the minimal social relations invariably associated with human life.

the dignity of rational, autonomous individuals each with a God-given right to it.

Clearly these issues could be taken further, but it is time to turn the discussion back to genomics and whether or not human dignity should be conferred on our genomes. For all their differences, individualistic and sociological accounts of human dignity are agreed on two key points. One is that dignity is assigned in the first instance to fully developed and differentiated human beings. The other is that the definitive offense against dignity is to treat whatever possesses it wholly instrumentally. On both accounts, accordingly, if genomes are imbued with human dignity, then unfettered genomic manipulation is an insult to it. But neither account makes it clear why human dignity should be assigned to genomes. Are there any resemblances or relationships between genomes and the functioning human beings that are the paradigm cases here, which might dispose us to speak of genomes as possessed of dignity? An analogy between fully functional human beings and minute blobs of DNA does not exactly force itself upon the mind. Even if we recognize an essential connection between ourselves and these blobs as a part of the natural order of things, why should such a connection imply that the blobs possess dignity?

These, of course, are currently much-debated issues and a variety of arguments, utilitarian arguments as well as the nonutilitarian arguments under discussion here, have been assembled on both sides. But neither the form of the debate nor the credibility of even the most abstract of these arguments can be understood without reference to an underlying distribution of preexisting intuitions and immediate responses. Much the most important of these is our tendency to respond to our fellow human beings as embodied creatures, inseparable from their physical or material manifestation. When we confer dignity upon them, their bodies are imbued with it and cease to be mere profane objects, and the consequent attitude of respect for the human body tends to be extended to whatever is continuous with it materially. The intuitive associations created in this way can have a very considerable force and intensity, as is apparent, for example, in the respect that is very widely accorded to corpses. Of course, how far these intuitions ought to be trusted and acted upon remains itself a matter for debate. Centuries-old arguments about whether the physical body is part of the very essence of the individual or merely a container for that essence (conceived of as spiritual or immaterial) continue to this day. And even if the body of a person is regarded as in some sense her material essence, it by no means follows that earlier stages of the life cycle materially continuous with the body of a person should be accorded the same status as that body.

Human dignity has been contentiously attributed to fetuses, embryos, and zygotes in recent years, as different stages of the life cycle have been characterized as, in essence, human beings; and in a further extension of the notion human dignity is now also being attributed to "the human genome." We could rightly object here, of course, that a genome can only ever be a part of an organism and never its entire material manifestation as a zygote may perhaps be; nonetheless, intuitions of material continuity do sometimes now extend not just to zygotes but beyond them to zygotic genomes, and they do prompt the thought in some of us that in these genomes reside the essences of complete human beings. Thus, far from being special cases, debates about how these genomes should be treated and how far it is permissible to modify them, to experiment upon them, or to destroy them along with the cells they inhabit, are now exemplary instances of the many similar debates that currently surround the treatment of human life at the level of the cell and below. Although we shall not dwell upon this, we are aware that the arguments we shall shortly discuss are also extended to zygotes, blastocysts, and embryos and deployed to rationalize the different kinds of treatment accorded to the germ cells, stem cells, and somatic cells of human beings.

It is hard to exaggerate the importance of intuitions of material continuity as the crucial structural underpinning of controversies of this kind. They are what ensure, for example, that debates about genomes and dignity focus on the germ line. It is on entities temporally continuous with the body of the mature individual, and intelligible to the mind as manifestations of the entire individual in a different form, that debate focuses. Nonetheless, it would be wrong to imagine that experience of material continuity suffices to fix the nature of these debates. The intellect needs more than this if it is to accept the validity of any given account of where the bodily existence of a human being begins and ends, and of the boundaries that demarcate it from nonliving matter. What verbal accounts need above all here is the support of authority. Historically, churches and religious institutions were long the recognized providers of this. And in the Christian tradition, and particularly the Roman Catholic tradition, they provided an authoritative picture, consistent with intuitions of material continuity, according to which the natural essence of the individual was present at almost every point of the life cycle. But today, it is widely agreed that empirical knowledge must be granted a major role in resolving such issues, and the natural sciences are now the recognized authorities on empirical matters.[8] Indeed, their authority in the empirical domain is now acknowledged by the Catholic Church itself and by most of the theologians and philosophers currently involved in these debates.

Human genomes and the order of things

Principled opposition to any manipulation of genomes in the cells of the human germ line is very common today, and such an uncompromising, "fundamentalist" stance can be rationalized in a number of different ways. A major strand in most such rationalizations, however, is the claim that we have here objects of a special sort that are not for us to meddle with. It is widely assumed that the zygote is in some sense a human being (just as an embryo is, and a fetus is, the critics often go on to say) or at least that it is latently or potentially a human being (just as a "totipotent" stem cell is, they might say), and that the genomes it contains are either what make it a human being, or what give it the capacity to become one. In a nutshell, the essence of a human individual is assumed to be present, in some sense, in these genomes. And while we may act upon and modify the body of an individual for her own good, so to act on her originating genomes is to transform her into someone else, or even into something else. And these are things we have no right to do.

   This position requires us to accept an essentialist conception of genomes and to confine ourselves entirely to an individualistic frame of thought; we have already set out our reservations about doing either of these things. But the position can also be called into question without attacking these basic elements. If the zygote is indeed a human being, should it not be *entitled* to modifications conducive to its good, just as other human beings are, rather than being "protected" from modification? Even if we accept an essentialist view of its genomes, may we not still regard it as legitimate to manipulate those genomes, provided we set clear bounds on what is ethically permissible? This is a widely supported view, defended at length by several authors, including Jürgen Habermas, whose book *The Future of Human Nature* (2003) we

8 This is not to say that the natural sciences have special standing in pronouncing on the presence or absence of essences. Authority extending over the empirical domain will not suffice for this. But if the issue is where there is and is not material continuity, or where there is or is not a particular kind of object customarily identified as being or carrying an essence, then the sciences can claim standing.

shall discuss in more detail in the next section. Like the fundamentalists, Habermas has an essentialist view of human genomes (p. 115), but he nonetheless refuses to support an unqualified prohibition on their manipulation. Instead, he draws a line between their repair and restoration, which in his view are defensible, indeed desirable, and their improvement and enhancement, which are not. Here, however, we come face to face with one of the main reasons why so many critics are reluctant to deviate from an uncompromising fundamentalist stance, in the shape of the notorious difficulties involved in making a distinction between these different kinds of genomic modification. These difficulties were clearly exposed in the debates prior to the Second World War on how to draw the line between negative and positive eugenics, although an intriguing inversion in our ethical orientations has since occurred. Then, it was negative eugenics—the elimination from the breeding population of those who were supposedly inferior or defective—that was most widely excoriated, and a positive eugenics that sought to improve the stock by encouraging the best to multiply that seemed more defensible. But today the elimination of heritable defects is far and away the most widely practiced and vigorously supported form of eugenics; and a eugenics of improvement, such as the intentional enhancement of genomes would now in effect constitute, is widely rejected as hubristic.

How are we to draw the line between repair and enhancement? And why do commentators from so many different standpoints agree that it needs to be drawn and that it marks the line between what is ethically reputable and what is not? Many critics of enhancement agree with the fundamentalist critics in accepting an individualistic ethical framework and an essentialist view of genomes. They oppose enhancement because it is not a mere improvement on the natural endowment of the individual, such as we routinely make by use of spectacles and hearing aids, but a transformation in her essential nature, and hence an illicit interference in the natural order. But here, in addition to essentialism and individualism, a third foundational assumption is discernible: that there is a natural order in the world, a condition of things that is at once how they are normally and naturally and how they ought to be morally, and that there is a normal-natural condition of our genomes that is a part of that order. And while an assumption of this sort is an option that frequently proves attractive to fundamentalists, it is much more than an option for those who would mark out a line between repair and enhancement: for them it is crucial. To repair something is to return it to its proper state, and to deploy the notion we need some conception of what that proper state is. As far as genomes are concerned, if we lack any sense of their "normal-natural state" or of their proper state as a part of the natural order, then we are free to identify what is proper just as it suits us and to do what we will in the name of repair or restoration.

Conceptions of natural order go back a long way, and the term "natural" here has not always meant what it does in our current scientific age. We tend now to make a stricter separation than our ancestors might have between conceptions of order in the empirically revealed world and conceptions of good order in a moral and ethical sense. Even so, our current notions of human health and well-being continue to run these two kinds of conception together, and this makes it easy for us to imagine that there could be a natural order at the level of the physical structure of the genome that is also the natural order from which healthy human beings result. A natural order of roughly this sort, identified with health and well-being, would indeed also be a moral or ethical order that would allow us to distinguish genomic manipulations that enhance from those that merely repair or restore. Nonetheless, any such ordering must still be identifiable as an empirical feature of the genome itself. And as we noted earlier, most of the philosophers and theologians involved with this kind of issue now accept that the natural sciences have become the legitimate authorities on empirical matters, and that their accounts of the natural world ought to be deferred to. In particular, although it has done so only very recently, the Roman Catholic Church now accepts that it should defer to the sciences here—and in the present context to biology. And since the Catholic Church also makes a moral distinction between the repair and the enhancement of genomes, it is interesting to ask how it proposes to identify their normal-natural state and what role it is willing to assign to the biological sciences in doing so.

In the Vatican Statement on Creation and Evolution (International Theological Commission, 2004),[9] a clear moral difference is asserted between the repair of human genomes, which in principle might be acceptable, and their enhancement, which is not:

> Germ line genetic engineering with a *therapeutic* goal in man would in itself be acceptable were it not for the fact that is it is hard to imagine how this could be achieved without disproportionate risks. (2004; para. 90)

9 The statement was published by the International Theological Commission under the presidency of the then Cardinal Ratzinger.

> The uniqueness of each human person . . . belongs intrinsically
> to him and cannot be instrumentalized in order to *improve* some
> of these characteristics. (2004; para. 91)

The statement also offers specific illustrations of the distinction, citing accepted biological and medical knowledge. This is most clearly apparent when the document refers positively to gene therapy.

> Gene therapy directed to the alleviation of congenital conditions like Down's syndrome, would certainly affect the identity of the person involved with regard to his appearance and mental gifts, but this modification would help the individual to give full expression to his real identity which is blocked by a defective gene. (para. 91)

It is true that the science actually cited in the statement establishes only a weak connection between its theological and ethical reflections on the one hand and scientific knowledge on the other, and that the cited material is always firmly subordinated to nonscientific schemes of interpretation.[10] Even so, the statement appears to concede that how things are genomically and genetically is something for empirical science to ascertain, and that the boundary between the repair and the enhancement of genomes must cross terrain mapped and described by biology. Moreover, there are other gestures to the sciences in the statement that signal respect not merely for their particular empirical findings but for their higher-level theoretical achievements as well. One example here is the recent explicit acknowledgment, repeated in the statement, that the theory of evolution (suitably understood) is "more than an hypothesis," but of even greater interest in the present context is the following:

> Thomistic anthropology . . . drawing upon the philosophy of Aristotle, understands body and soul as the material and spiritual principles of a single human being. It may be noted that this account is not incompatible with present-day scientific insights. Modern physics has demonstrated that matter in its most elementary particles is purely potential and possesses no tendency toward organization. But the level of organization in the universe,

10 Although Down syndrome is not a condition caused by a defective gene, in the current context this can be passed over as an insignificant point.

which contains highly organized forms of living and non-living entities, implies the presence of some "information." This line of reasoning suggests a partial analogy between the Aristotelian concept of substantial form and the modern scientific notion of "information." Thus, for example, the DNA of the chromosomes contains the information necessary for matter to be organized according to what is typical of a certain species or individual. Analogically, the substantial form provides to prime matter the information it needs to be organized in a particular way. This analogy should be taken with due caution because metaphysical and spiritual concepts cannot be simply compared with material, biological data. (2004; para. 30)

On this analogy, particular DNA sequences organize inert matter into a living creature with a given species nature, and we may also presume, given paragraphs 90 and 91, that "defects" in the DNA sequences will organize matter to form "defective" creatures. Scientifically authoritative "information," revealed by the great genome-sequencing projects, also provides morally salient insight into our natural essence that will underpin a demarcation between the repair and enhancement of genomes. However, the Vatican statement also urges caution in the use of this analogy and rightly emphasizes that "metaphysical and spiritual concepts cannot be simply compared with material, biological data." And indeed this criticism is so powerful that it calls the value of the whole analogy into question. We should not look to the results of the Human Genome Project (HGP) in the hope of a revelation of the natural order in human DNA. All that these great projects provide is so many sequences of particular lengths of DNA, run together by technical devices, and referred to as "the human genome" by practitioners intent upon coordination of and around a range of esoteric purposes (Bostanci, 2008). There is no natural order identifiable here, only a conventional ordering of "material biological data." There is no way in which this will serve as an acceptable basis on which to distinguish the repair of human genomes from their enhancement, and no way in which the further processing of data by biologists for their purposes will automatically make it authoritative for moral and ethical purposes. But the way in which the science is addressed conveys the false impression that this data will so serve, even if its employment for these latter purposes is no simple matter.

Difficult issues of great importance arise here about the domains of epistemic authority of different institutions and the relationships between them, but we shall limit ourselves to discussing the knowledge and authority of the biological sciences and how those who accept it should seek to relate to it. In particular, we shall argue that the Vatican statement, and by extension the religious authorities speaking through it, fail to engage appropriately with the biological sciences, given that they recognize the authority of those sciences on empirical matters. It is noteworthy here how scant and selective are its references to specific empirical findings, the "material, biological data" mentioned by the statement, but this is not our main criticism. Nor do we object to its citation of the informatic account of genomes, even though we have pointed to its limitations in previous chapters and not all scientists are content with it. The informatic account has considerable standing in science and those who take science on authority can scarcely be criticized if they thereby accept an unsatisfactory theory. The point of departure for our argument is the *way* in which the statement engages with the theory and the "data," and particularly with the former.

Our suggestion is that in engaging with science only at a very general level, through theory, abstract concepts, and idealized accounts, the statement fails properly to engage with it at all. We need to recognize that even if science sometimes appears to describe the essences of things, such descriptions are very far from being the essence of science. Essences and abstractions and the general theories that refer to them are often misunderstood as the summation of scientific knowledge, but they are better thought of as devices used to coordinate understanding and activity within science and to link and organize empirical findings. If one engages with the sciences wholly at the grand theoretical level without addressing particular findings—as equally, if one engages only with particular findings without reference to theory—missing the level where theory and findings run together in the *doing* of the science, one denies oneself access to the most trustworthy and authoritative elements of scientific knowledge. All that can then be accessed is material like that in the Vatican statement, which ripped from its context implies nothing beyond itself and can be readily interpreted and elaborated for whatever extrinsic purposes are expedient.

Anyone who genuinely accepts the authority of the natural sciences on empirical matters needs to recognize where that authority is located and how it is best addressed. The authority in question is communally based authority, most strongly imparted to knowledge that is collectively generated and scrutinized, and put to the test through shared use. The knowledge imbued with the greatest scientific authority is not grand theory but mundane, routinely accepted knowledge bound up with scientific practice and immediately surrounding networks of belief. It is in the course of practice, generally the practice of scientific research, that the distinctively valuable empirical knowledge of the sciences is laid down, semantically stabilized, transmitted, and, crucially, constantly evaluated and reevaluated, so that it is indeed the *authoritative* knowledge of the scientific field.[11] And it is only when situated in contexts of ongoing scientific practice that highly theoretical forms of discourse, for example discourses referring to "the human genome," may be employed with proper awareness of their limitations and with little danger of being misled by unduly realistic understandings of them. Sadly, however, few besides scientists themselves find it easy routinely to deploy scientific knowledge in this way and draw reliable conclusions from it. For others, engagement entirely at the level of general theory, about abstract essences and ideal objects, may be the preferred option, and all the more so because the results are typically so much more amenable to further interpretation and elaboration. But difficult though it is to engage adequately with science as an authoritative source of knowledge, this neither justifies inadequate forms of engagement nor validates their outcomes.

What would be the result of proper engagement with biology in the present context? Whatever else, it would be strikingly different from what is found in the Vatican statement. A number of interdependent components of the "working knowledge" of the field would prove hard to reconcile with an essentialist understanding of genomes and the existence of a "natural order" in their DNA sequences. Think, for example, of the wholly uncontroversial fact that genomic DNA is constantly changing, whether in the course of normal bodily processes, or from "errors" in the course of replication and recombination, or because of mutations induced chemically, or by radiation. The long-term effect of such changes considered in isolation has to be a loss of order; as time passed, the "interesting" order in the DNA would disappear and the corresponding life forms along with it. Order persists, if it persists at all, only as a dynamic equilibrium wherein disorder-creating processes are opposed by order-creating ones. At the level of the cell, normal functioning itself involves the invisible monitoring and constant "repair and restoration" of DNA sequences.[12]

---

11 This harks back to the themes discussed in chapter 1, and arguments there attributed to Rheinberger (1997) and Kuhn (1970, 1977; and see also Nickles, 2003), although they can of course be traced back much further, to writers in the pragmatist tradition and earlier.

12 It is worth noting, lest anyone be tempted to an Aristotelian view of these monitoring and repair processes and their *telos*, that they will sustain all kinds

At the level of the life cycle, the heritable mutations and variations that continually occur usually result in "defective" organisms that survive and reproduce less often than "normal" forms and tend accordingly to be continually eliminated. If the rate of either production or elimination should change, as can occur for all kinds of reasons, the system moves to a new equilibrium.

Differential removal of "defective" variants begins quickly in humans. They are heavily overrepresented in the spontaneous abortions and miscarriages that are the fate of a very large proportion of fertilized ova. And natural selection continues to have it in for them subsequently, whether because of their greater vulnerability to disease or because of their lower fertility and reduced likelihood of mating. But part of what determines their fate is human agency, and today it is an ever-increasing part. Selective mating, and selective abortion intentionally undertaken, contribute to the elimination of "defective" variants,[13] whereas determined efforts to extend the life span of human beings, and particularly of those whose inheritance might otherwise consign them to an unduly rapid demise, have effects in the other direction. Indeed, the ever-increasing success of efforts to sustain life in prematurely born infants and the continual reduction of the period from conception to viability is already resulting in the survival of more and more individuals with genetic/genomic "defects," and hence in the accumulation of more and more "deleterious" variants in the stock of DNA.

None of this discussion is intended to justify unrestricted genomic manipulation or an unapologetic eugenics of improvement. We are merely pointing to themes that tend to be neglected in ethical debate, in part because those involved are reluctant to engage with science at a level that might unduly constrain what they can make of its knowledge. With such engagement their vision, and hence that of their audiences and followers, would have been a more complex and a more detailed one, but this is not always what is wanted when ethics and policy are being debated. Certainly the details passed over in this case, relating as they do to genomic change and variation and to what these imply at the collective or population level, are hard to square with the existence of a

---

of sequences, and that their presence in cancerous and other unwanted cell types makes them part of what constitutes some diseases.

13 The "defective" variants eliminated by intentional abortion are, of course, not always the same as the unwanted variants also eliminated in immense numbers by the same means, among which are a large proportion aborted because they are female.

"natural order" in "the human genome" and hard to evaluate without going beyond the confines of the individualism that increasingly frames the debate. Of course, those whose duty it is to establish ethical standards and give them political expression do not have to be swayed by these neglected aspects of the science, but they do have a responsibility to address them. Indeed, such is their significance that to ignore them is tantamount to scorning the significance of empirical biological science altogether.

The themes and findings we have mentioned are clearly relevant to any assessment of the impact of a total ban on the manipulation of human genomes, and they create difficulties for efforts to rationalize the ban with the claim that manipulation disturbs the natural order of things. Let us confine the discussion to human health and well-being: if we cease to act in ways that disadvantage deleterious genomic variations, whether through discontinuing the selective terminations currently permitted or through abandoning attempts to find ways of "repairing" them, these variations will accumulate in the population and re-equilibrate at a higher level of incidence. If at the same time we persist in our efforts to preserve all human life, and particularly to bring every implanted embryo to term, that level of incidence will increase. If we continue to facilitate reproduction as an individual human right, it will increase further. If technologies that allow more and more of us not just to survive but to multiply are energetically developed and applied, it will increase still more. Where is the "natural order" here, where all kinds of dynamic equilibria between different genetic/genomic variants may exist and we ourselves have long appeared on both sides of the equations that determine their relative frequencies, at once encouraging and discouraging the accumulation of "deleterious" genomic variants? We can scarcely look back in this particular case to the "natural order" that existed prior to our own emergence, but we cannot look any later either, for as long as we have been around we have been disturbing whatever order surrounds us. It is tempting to suggest here that original sin dates back to the time when the twitch of a human finger first disturbed "the natural order" and that in our subsequent fallen condition if we prohibit disturbances of one sort we merely expose and amplify disturbances of another sort.

Even if fundamentalist opposition to genomic manipulation cannot be rationalized as protection of the natural order of things, it may perhaps be rationalized in other ways. Opposition not to repair but solely to enhancement, on the other hand, cannot adequately specify what it is opposed to in the first place, if there is no natural order to act as a dividing line. Consider the question of whether an aged body is in a normal-natural state or one that would benefit from restoration. Is the "repair of defects" in human genomes justified to prevent physical processes of aging, or is this merely a misleading description of genomic enhancement? A very great deal could hang upon how this question is answered, yet there is no revelation of natural order in "the genome" to serve as the basis of an agreed answer here and no alternative route to one apparent either.

Human genomes and the dignity of human life

There is no scientific warrant for belief in a "natural order" in the DNA sequences of our genomes, or for the idea that our natural essence resides at the same address. And accordingly there is no scientific basis for those arguments against genomic manipulation that identify (zygotic) genomes as possessed of human dignity. But even if this inclines us to dismiss such arguments before they even get started, it can be interesting to look at how they are designed and presented. Although scientists are often (rightly) suspicious of them, the philosophers and theologians who make these arguments contribute to the processes whereby society and its institutions come to terms with scientific insights and innovations, and they are currently playing a part in bringing genomics into the ambit of law and regulation. There are many sources we could turn to here, but we shall focus on arguments citing dignity in the short book by Jürgen Habermas (2003) that we referred to earlier. These arguments are at once distinctively philosophical and closely coupled to the political, legal, and regulatory issues surrounding their subject. They make close reference, for example, to the Articles of the German Constitution and the Charter of Basic Rights of the European Union. They are closely bound up with critical commentary on the political context in which a new eugenics is evidently on the point of emerging, and with further arguments emphasizing the dangers its uncontrolled proliferation represents in political economies dominated by the market and rampant consumerism. And finally, they are the arguments of someone whose words have long commanded widespread respect and had a proven influence in areas of life and institutional settings extending far beyond the context of academic debate.

Before turning to his actual arguments, it is interesting to look first at Habermas's account of *why* he is going to argue as he does. There is, he says, an argument that concludes that the fertilized egg cells (zygotes) of humans possess "in the strict sense human dignity" (p. 37). On this argument they ought to be accorded the status of human subjects with full human rights. Habermas confesses that he is motivated to use this argument, since anything less would "leave the door open a crack for an instrumentalisation of human life" (p. 38). And he clearly implies that, mindful of the "provocations" of genetic engineering (p. 42), and of the appalling prospect of an untrameled "genetic marketplace" arising in our current consumerist societies (p. 75), he personally would happily accord complete protection to the zygote and its genomes in order to prevent their being treated instrumentally. Nonetheless, Habermas does not take forward this argument. He argues instead that nonpersonal manifestations of the human should be recognized as possessing not full "human dignity" but the "dignity of human life," a lesser dignity yet one that still requires them to be afforded some protection against in-strumental manipulation (pp. 29ff). The reason he gives for advancing this modest proposal is that he sees no chance of securing a "universal consensus" for the more ambitious one. In this context he regards it as crucial to secure a consensus extending across all those diverse groups of people with quite different worldviews who constitute the citizenry of a modern state. That there should be a universal consensus here is actu-ally more important than what precisely the consensus consists in, and because of this Habermas takes up the argument he believes to have the greater persuasive power in the specific social context with which he is concerned. We shall return to discuss this important point in the next section of the chapter.

Not everybody can be persuaded to accept that embryos or zygotes or genomes possess human dignity in the full sense. But why might ev-erybody be persuadable that these objects have dignity as nonpersonal forms of human life? Habermas responds to the question in two ways. First, he notes that all human beings, whatever their culture or world-view, do *in fact* recognize that nonpersonal human life merits special treatment. Always and everywhere, corpses are accorded a special sig-nificance, for example, and other forms of human material are treated as more than mere material, suggesting that the universal consensus he seeks may in a sense be there already, lying latent. Second, he identifies arguments that ought to move the members of any culture at all, because they concern the species as a whole and what he calls its "ethical self-understanding." In particular, he argues that "as soon as adults treat the desirable genetic traits of their descendants as a product they can shape according to a design of their own liking, they are exercising a kind of control over their genetically manipulated offspring that intervenes in the somatic bases of another person's relation-to-self and ethical freedom" (p. 13). Habermas refers to this deliberate shaping by genomic manipulation as "alien determination," and the argument against alien determination dominates much of the discussion in his book.

Despite his concern with the "self-understanding of the species," however, Habermas actually develops this central argument at the individual level. There is, he says, a program in the genome, varying from individual to individual, that determines the major physical characteristics of the subsequent person and something of her behavioral and mental characteristics as well. Normally, the specific program is an accident of nature. But if it is intentionally altered or added to, whoever is responsible is participating in the design of another person, who may in due course come to know that she has been programmed or designed by someone else. Here, we are told, lies the tension with our ethical self-understanding as it is currently constituted. Our present ideal ethical vision of ourselves is as free and equal human beings, "born not made" as Habermas puts it, all of whom have the same standing in their social relationships, all of whom are reciprocally accountable for their actions. But where a human being has been deliberately designed and is not linked to the designer by the genuine reciprocity between one freeborn being and another, this form of ethical self-understanding is not available to her and the result may be that she suffers serious psychological harm. Imagine, for example, that a super-intelligent baby is desired.[14] The deliberate reprogramming of a genome to produce such a child would be an instance of what Habermas calls "alien determination," and would, he suggests, be liable to engender psychological trauma in the child for that reason alone.[15]

14 This is the sort of thing that parents are often said to aspire to do, and Habermas does evince an intense concern with the individual parent as consumer, in need of collective control through the regulatory agencies of the state if she is not to act irresponsibly in a genetic marketplace (2003, p. 75). It is intriguing that while taking an obsessive interest in the pathologies of the market in this context, Habermas says next to nothing about the possible consequences of state irresponsibility.

15 To those familiar with Habermas's earlier work where the opposition to instrumental actions and the power of "system" is identified as residing in community and lifeworld, the representation of the harms of alien determination in psychological terms and with reference to the relationship of two abstract individuals will surely be surprising.

There are some striking flaws in Habermas's case against alien determination that we need to emphasize. First of all, the threat to the "ethical self-understanding of the species" posed by alien determination is more or less equated with the psychological harm that will be done to "reprogrammed" individuals. What if there was no psychological harm? Habermas is given to making empirical claims without the backing of any empirical evidence, and this is a striking instance on which a considerable part of his entire argument is made to hang. Of course it remains possible empirically that psychological harm could be done to those who become aware that they are "made not born," but fortuitous empirical confirmation of this sort, rather than rescuing Habermas's case, would merely bring us face to face with another of its weaknesses.

Habermas's argument is formally defective as well as empirically unsubstantiated. Alien determination is condemned as inconsistent with the ethical self-understanding of the species and as an affront to "the dignity of human life." But the actual argument offered to support this judgment looks completely beyond the "reprogrammed" genome to the adverse consequences for the subsequent person, liable to suffer psychological trauma through her sense of her *human dignity* having been compromised. It is a utilitarian argument, or at least a consequentialist one, and whatever power it has derives from our concern for an actual person and *not* from our concern for the nonpersonal human material in the zygote. Any concern for *the dignity of human life* is gratuitous here. Habermas's argument, if accepted, justifies a ban on the manipulation of zygotic genomes only insofar as they are going to become self-aware human beings, and then regardless of whether or not they themselves have any special dignity—which is perhaps as well given our indifference to the innumerable fertilized eggs that pass into the sewage system as a consequence of spontaneous abortions.

It is worth reiterating at this point that for all its fulminations against the manipulation of human genomes, Habermas's text is not opposed to it. It argues, or rather it accepts, that the deliberate "reprogramming" of the human genome can be desirable where repair rather than enhancement is involved. But the key argument from alien determination does not discriminate between these two kinds of "reprogramming"; it counts against deliberate "reprogramming" of all kinds. There is a major inconsistency here that Habermas eventually appears to recognize. But in responding to it and "rescuing" the ethical distinction between repair and enhancement to which he is committed, "the dignity of human life" is once again treated as superfluous. In a major shift of position quite late in the text, Habermas proposes that deliberate genomic "reprogramming" should be assessed by reference to the *attitude* that informs it. "Reprogramming" informed by the instrumental attitude of the technician is never justified, he tells us, but where it is informed by the attitude of the healer it may be. The appropriate attitude here will involve an intention only to restore or repair and not to enhance or improve, an acceptance that the line between restoration and enhancement must not be determined by individual judgment but by reference to collectively agreed understandings, and a wish to act only in ways to which the subsequent individual would give informed consent. On these criteria the reprogramming of a genome may be permissible, even though it is an instance of the deliberate redesign of one individual by another.

The criteria of good ethical practice cited here are, of course, those that currently inform the relationship of a doctor with her patient. And what they evince concern for, beyond the overall responsibility of the practitioner to the collective, is the welfare and the dignity of the subsequent individual. It is our shared sense of the normal condition of human individuals that allows the notion of healing to be applied, here or elsewhere, and permits a distinction between repair and enhancement to be made. And it is respect for the *human dignity* of normal human individuals that obliges us to imagine what "repairs" to their genomes they might retrospectively give informed consent to. Thus, Habermas's revised argument takes no more account of the *dignity of human life* than did his previous argument from alien determination. As before, the only kind of dignity actually salient in the evaluation of ethical conduct is human dignity in the full sense, the dignity conferred upon each other by fully formed human beings and primarily expressed in their social relations with each other. Although Habermas himself does not emphasize it, the actual endpoint of his discussion is that ethical conduct in relation to genomes should be evaluated entirely by reference to subsequent individuals.

It is worth turning from Habermas's suspect arguments here, to reflect for a moment on the merits of this mode of evaluation, which are considerable. If the appropriate treatment of human (zygotic) genomes is inferred entirely by looking to the welfare and the dignity of subsequent individuals, then on the one hand a whole array of possible interventions currently abhorred are confirmed as perverse and misconceived and the prohibitions set upon them are thereby further justified. This can only be to the good, even if distaste for the deliberate production of luminous children, or handicapped children, or docile children scarcely stands in need of further justification. On the other hand, and equally to the good, challenging questions are put to some aspects of our current practice and dominant modes of thought.

Let us cite just two of these. First of all, arguments from the subsequent individual have no application where there is no subsequent individual. It becomes wholly acceptable to manipulate human genomes in zygotes and aggregates of cells provided they are disposed of afterwards. To put it another way, scientific experimentation involving any kind of manipulation of the genomes of nonpersonal forms of human life, whether in zygotes, or stem cells, or cell nuclei transferred to the eggs of other species, or whatever else may be so defined, becomes entirely legitimate provided only that no subsequent individuals ensue.[16] A great number of currently contentious practical issues might quickly and sensibly be resolved if this mode of evaluation were generally accepted. Second, concern for the subsequent individual does not just feed back via the fetus and the embryo to zygotic genomes; it feeds back to the genomes of the gametes as well and possibly further still. The point where two gametes become one organism now ceases to be of any particular ethical importance, and the arbitrariness of the assumption that a new human life begins with the fertilized egg is cruelly exposed. There is much to be said for highlighting the rarely questioned distinction between the sacred zygote and the profane gametes. The prospect of the deliberate uncontrolled manipulation of zygotic genomes that may eventually become the genomes of self-aware human beings ought surely to appall us. But the genomes of gametes may become the genomes of zygotes, and whatever appalls in the one case should also appall in the other if our concern lies entirely with the subsequent individual. Thus, as a basis for an ethical evaluation of how we may manipulate nonpersonal human life, our concern for human dignity in its paradigmatic sense may diminish as well as expand the realm of what is considered legitimate.

Arguments and institutions again

Let us return now to Habermas's arguments and briefly retrace the path they take. Habermas begins by expressing sympathy for the view that human zygotes and their genomes possess human dignity in the full sense and should be treated as surrogate human beings. This position is then set aside, and it is suggested that in the interests of securing a wider consensus the fertilized ovum and the genomes within should be regarded as

---

16 Of course the problem remains of what is a nonpersonal form of human life and what is not, and as the abortion debates illustrate there are many societies in which the achievement of a consensus on this is nowhere on the horizon at present.

nonpersonal forms of human life possessed only of a lesser dignity, "the dignity of human life." Habermas then goes on to oppose the deliberate "reprogramming" of these genomes in order to enhance or improve the subsequent individual. This "alien determination" of human nature is denounced as a threat to the "self-understanding of the species" that is liable to give rise to psychological trauma in anyone, in any culture, aware that she is a product of it. But what this argument has to do with the dignity of nonpersonal human life, as opposed to the human dignity of the subsequently traumatized person, is never made clear. And neither is it shown why the argument does not apply to repair, which Habermas consistently wants to permit, as much as to enhancement. Finally, having apparently recognized this last problem, Habermas belatedly moves to a different line of argument and proposes that manipulation be permitted only where it is informed by the correct attitude, the attitude of the healer to a patient. This, however, merely further confirms the irrelevance of the notion of "the dignity of human life" and returns us to a concern with human dignity in the full sense as the basis for an ethical evaluation of the "reprogramming" of human genomes.

We have been strongly critical of Habermas's account. It relies on empirical speculations for which no empirical evidence at all is offered, and considered as formal philosophical reasoning, its key arguments are incoherent. There is also a vacillation between different arguments as the book proceeds that makes a striking contrast with the unwavering consistency of what is argued for. Nonetheless, the purpose of our discussion has not been solely, or even mainly, to offer criticism and expose flaws. Even if the flaws in the arguments were rectified and the entire text revised to conform to the highest standards of philosophical rigor, they would still not compel assent through the sheer force of reason. Like all arguments on moral and ethical issues both nonutilitarian and utilitarian, they would remain less than rationally compelling and liable to be opposed by other lines of argument no less reputable and well-constructed in a formal sense. A broader kind of critical appreciation is needed here, which takes account of the context of the argument. In that context, Habermas is evidently seeking to rationalize modes of conduct: he presents arguments likely to dispose us to act in a given way, leaving others to construct arguments likely to move us the other way. This is typical of how we engage discursively with controversial issues. Opponents argue from different precedents, constructing analogies that help their cases and downplaying others, and this is particularly apparent in controversies about how much of the dignity accorded to fully formed human beings should be extended to embryos, zygotes, and genomes.

Argument by analogy is designed to move opinion through reason, which in a sense it may on occasion do, but it cannot move opinion through the compelling power of reason since analogies are not compelling.

All this suggests that reflection on Habermas's arguments should begin by asking what he is trying to bring about as he deploys them. Of course this must be a matter of interpretation: the goals and purposes of an author cannot be read directly out of a text, and different readings are certainly possible in this case. If we examine what is explicitly argued for, then an acceptance of the controlled use of genomic manipulation in the cause of a limited form of eugenics is clearly what is being sought. But there are good grounds for believing that this is not the primary objective. Habermas's himself explicitly states that a broad agreement on one specific policy for the regulation of genomic manipulation is more important than the specific policy that is agreed on. His own words imply that his primary purpose here is that of consensus formation and that the specific agenda around which he is striving to secure a consensus has been selected with that larger aim in mind.

Habermas has often spoken of an ideal universal consensus, to be secured through reflection and rational discursive engagement, but it seems that a slightly more specific and less Utopian consensus is actually being sought in this case. The text emphasizes both the importance and the sheer difficulty of consensus formation in institutionally complex and culturally diverse societies in which many different worldviews exist. And it specifically refers to "a clash between the spokespersons of institutionalized science and those of the churches, engendered by the advent of genetic engineering" (2003, pp. 101ff.). In both the more abstract philosophical arguments of the book and its reflections on what should be the response to genetic engineering in the context of law, politics, and regulation, these are the two constituencies above all that are invited to agree with its conclusions. And indeed there is much else to suggest that its primary goal is to promote a specific kind of consensus reconciling, in some sense, "institutionalized science" and "the churches," that this is the end toward which the arguments are designed to lead, the end from which their construction begins.[17]

17 The nature of the proposed consensus is a side issue here, but its effect would seem to be to allow scientists all the freedom they might desire for their research on nonpersonal forms of human life. It is intriguing sociologically that while the text offers the churches an unrelenting respect and deference in striking contrast to the withering criticism it frequently directs at institutionalized science, it is the latter that would be the clear beneficiary of Habermas's actual proposals.

It is tempting to be dismissive of arguments constructed in this strategic fashion, beginning with the desired conclusion and working backwards to the premises. And our earlier discussion of some arguments of this sort in chapter 5 may well have reinforced the view that there is little to be said for them. But rationalizing arguments, as we shall henceforth call them, are neither intrinsically flawed nor inherently objectionable, and if we reflect on how they function in general terms, it can lead us to a much more positive evaluation of them as a genre. A good way to a general understanding of how rationalizing arguments actually function is to imagine where we would stand without them. New phenomena and new technologies create the need for changes in shared understandings and routinized collective responses. When they impinge upon us they typically set in train a movement in how they are addressed, from individual perceptions to collective representations, from intuitive evaluations to shared judgments, from immediate particular reactions to routinized and coordinated responses. At the institutional level, the initially novel states of affairs must become familiar objects of scientific knowledge, religious and moral doctrine, and legal regulation, not to mention of everyday understanding. Crucially, at the level of behavior, there must be a movement toward coordination and a renewed or extended agreement in the practice of the society. This movement, which Habermas himself has discussed in some of his most impressive writing, is effected by communicative interaction, in which rationalizing arguments may feature prominently. And in culturally and institutionally differentiated societies like our own, specialists in the genre may be major suppliers of such arguments.

Unfortunately, little has been written on the evaluation of rationalizing arguments. We have a strong sense intuitively that some have more merit than others, irrespective of their conclusions, but it is important to recognize that much of this sense derives from considerations external to the arguments themselves.[18] First of all, it derives from the functions the arguments fulfill, which are often not merely extrinsic to them but unrecognized by those who advance them. Their coordinating functions

18 In focusing on the externalities as we do here, we do not want to create the impression that arguments, or more precisely materials offered as arguments, should be judged wholly and entirely in terms of effects in particular contexts. If such materials deceive and mislead, for example, then of course they merit criticism for doing so: the last things we would wish to defend are arguments like those employed by prominent politicians prior to the attack on Iraq by the USA and the UK in 2003.

in particular are usually left implicit, and only rarely explicitly referred to by those making the arguments in the way that Habermas does. Typically, we coordinate around new things by arguing about the things, not about ourselves and how we ought to be coordinated. To coordinate around our treatment of zygotes and their genomes, for example, we talk about how they are, not about how we are. Second, rationalizing argument is a form of social activity. It arises in debates in which a whole range and variety of arguments are collectively produced, no one of which can be evaluated independently of the others, and all of which are part of the context in relation to which the others are evaluated. Finally, since rationalizing arguments are generally arguments by precedent or analogy, the sense that they are valid must depend on how susceptible given audiences are to the precedents or analogies involved. To understand the evident power of the arguments it is necessary to look to the audiences that are moved by them. The analogies the arguments invoke may remind audiences of what they already believe, for example, or resonate in some way with existing beliefs and intuitions, or give voice to them in ways that the audiences themselves may previously have not been able to. We have already described how Habermas identified "the dignity of human life" as something already implicitly recognized everywhere and expressed in the practices of all cultures, and how he fastened upon the notion accordingly as something around which to attempt to mobilize agreement and secure consensus.

It is moot how sound Habermas's judgment was on this particular matter, just as it is moot how far his intervention in this context will contribute to the creation of the kind of consensus he is seeking. But in highly differentiated multicultural societies like our own the kinds of argument his work exemplifies may still have valuable coordinating functions even if they come nowhere near to engendering an extended consensus. Examination of the remarkably high level of coordination achieved in these societies indicates that its final, vital layer is often the result of the systematic organization of disagreement between internally coordinated subgroups: it is the product of an agreement to disagree, as it were, and in how to disagree, rather than of a consensus in any more profound sense. And the debate that work such as Habermas's both constitutes and provokes has a role in identifying where we agree and disagree and which disagreements we may hope to overcome and which not, permitting the great sacred organizing principle evident in all these "advanced" societies to be applied: make what is agreed upon the basis of life in the public realm and the sphere of the state; consign what cannot be agreed on to the private realm and the sphere of the family.[19] Here is an organizing strategy that has again and again proved magnificently successful in creating order and coordination, not to mention peace and security, out of conflict and dispute, although of course there remain even now minorities guilty of what in these societies is the greatest sin of all, the sin of acting rightly according to their own "private" moral and ethical standards even in the public realm, fundamentalism as it is commonly called.

Jürgen Habermas is recognized and honored in all the "advanced" societies as among the world's leading public intellectuals, indeed as one of those who helped to define the role of the public intellectual and make the case for its importance in encouraging a more participatory democracy. His interventions into political and ethical debates, however, once clearly directed to extending the range and intensity of public debate and coupling it to political action via new lifeworld-based social movements, have tended to become less radical over the years. In *The Future of Human Nature* this shift is so marked that it ought perhaps to cause disquiet to those promoting more participation in politics and a greater public engagement with science. For here Habermas seems in the main to speak to power and to seek a reconciling consensus between two of the great established institutions that dispense it. It is the spokespersons and representatives of science and the churches, especially the Roman Catholic Church, that have set the context for his rationalizing arguments. As to ordinary members of the public, they figure prominently neither as an implied audience nor as a topic of the text, in which they appear only as potentially irresponsible individual consumers, a potential market for genomic modification in dire need of regulation and control.

If it is indeed the case that Habermas now expects less of ordinary members of society and of participatory democracy, and has become more inclined to seek influence through engagement with powerful institutions, then he has moved in the opposite direction from the general trend in this area, particularly where issues involving scientific research and the implementation of new technologies are concerned. Certainly

19 Very often, of course, it is religion that is identified in differentiated Western societies as the area where agreement between cultures is not to be looked for, the area that has to be privatized, and science that is seen as the great hope, the body of empirical knowledge on which all may agree, even the key basis for a secular polity. For discussions of this focused on science, see Shapin and Schaffer (1985) and Barnes (2000, p. 125ff).

in the English-speaking world, the promotion of increased political participation by an active, knowledgeable, and empowered citizenry, even at the expense of a lessened credibility for scientific expertise, continues to build on its already extensive support, both as an intrinsic political good and as a possible way of solving perceived problems of trust in existing modes of governance and regulation.[20] But if there is indeed a major difference of opinions lying latent here, it is far from clear which is the more plausible.[21] Few would dispute today that how a society responds to scientific and technological changes is everyone's business, since the lives of everyone are affected by them. But the question of whether it is better for everyone actively to involve themselves in fashioning appropriate responses, or for elected representatives of "the public" and technical specialists in whom it trusts to take on the task, remains unresolved, as does the related question, not always distinguished as it should be from the first, of how epistemic authority should be apportioned between experts on "narrowly technical" issues, nonexperts with relevant knowledge and experience, and other people legitimately involved in the relevant decision making, up to and including the public as a whole.

The search for answers to these questions is currently a major preoccupation because it is relevant to our reception not just of genomics but of every kind of scientific and technological innovation. And since there are no indefeasible answers to these questions, and moral and political predilections will inevitably bear upon which proposals find favor where, we should take for granted the need for unending debate

20 See Wilsdon and Willis (2004) for what is currently the most uncompromising version of the participatory perspective in the UK. This advocates "upstream engagement," wherein "the public" involves itself in deciding not just how to use new science and technology, but whether and in what way it should be pursued and developed. The authors are not unaware of how their program may be supported both by some scientists and politicians as a way of fixing the public and by some campaigning radicals and activist components of the public as a way of fixing science (see pp. 19–21).

21 Indeed, the tensions here are no longer solely between advocates and opponents of the move from representative to participatory forms of democracy. Other versions of democracy, for example "deliberative democracy," have been suggested as possibly preferable to both, while argument continues among its advocates as to just what participatory democracy ought to involve. Compare the following from a far more extensive literature: Rayner and Malone (1998); Cooke and Kotheri (2001); Jasanoff (2003, 2005); Nowotney (2003); Pellizzoni (2003); Rayner (2003); Rowe and Frewer (2004); Mutz (2006).

around them. Whenever societies alienate powers and responsibilities to specialists and create a system of divided technical and scientific labor, as ours have done, they create dilemmas for themselves that they have to continue to confront. The greater the effort put into control of and/or engagement with scientific expertise, the less the efficiency gains from the division of labor become; complete control and engagement entail complete knowledge of what experts know, which implies no efficiency gains and indeed no experts. On the other hand, to disengage completely, and simply accept expertise passively, implies what is surely an undue level of dependence, as well as inviting corruption and depriving scientists themselves of epistemically valuable inputs from the wider context.[22]

Whatever overall modus vivendi is established here, specific problems will continue to arise. Precisely where the boundaries of different domains of expert authority lie can always be disputed in any given case, and the standing of specific scientists, even when giving testimony on their own research or on the use of their own technology, can always be questioned (Gieryn, 1999; Jasanoff, 2003, 2004, 2005). The commonest manifestations of this kind of contestation of epistemic authority have actually tended to arise between scientists themselves, especially those in different fields and/or employed by opposed sides as expert advisors on technical, legal, or policy issues, although disputes along the boundaries of science itself have been far from infrequent, and if participatory impulses continue to intensify will surely remain so. Considered from this perspective, Habermas's work serves to remind us that the institutional boundary between science and religion remains a contested one, and that genomics and associated areas stand among those fields whose domain of epistemic and ontological authority is liable to be challenged, on occasion, by theologians, whose references to sanctity and dignity cannot be directly opposed by scientific arguments. But as far as the resulting disputes themselves are concerned, there is nothing that sets them apart from other more mundane boundary disputes involving scientists or scientific fields, whether in how evidence is mobilized and interpreted, or how rationalizing argument is constructed, or how claims to authority are articulated and called into question.

22 Famous among studies showing the valuable contributions to technical debates that involved nonspecialists can make at the epistemic level are Epstein (1995, 1996) and Wynne (1989).

8

Conclusions

Genomics as power again

The time has now come to look back on the discussion as a whole and highlight its most significant conclusions. In our introduction, we stressed the need to run together two perspectives on genomics often kept distinct and separate. Genomics is at once a body of knowledge and a technology; it is a culture carried by specialists who both know things and do things, and the knowing and doing cannot be understood independently of each other. The point is of fundamental importance because without a sense of the field both as knowledge and as activity we shall be ill-equipped to grasp and respond to the problems and opportunities that will continue to emerge from it. Idealized visions of scientific fields, which present their knowledge abstractly as a revelation of how the world really is, may inspire the imagination in one way but they impoverish it in another. As beautiful as some of them may be, they are static visions, in which the human activities that are the springs of change of knowledge lie hidden, and of course not all of these idealized cosmological visions are even beautiful. We have tried to move beyond this kind of work here, in the direction of a less abstract and more naturalistically adequate general account of genomics and its associated sciences.

Culture abhors a vacuum. If everyday understanding of the sciences is not based upon the naturalism characteristic of those sciences themselves, then it will be based upon a nonnaturalistic or even an antinaturalistic perspective. For anyone who shares our own commitment to naturalistic understanding, this danger should suffice both to explain and to justify the kind of account we have tried to produce. But the need for a naturalistic understanding of genetics and now genomics is particularly acute because of the longstanding tendency to assume that they are in some sense exceptional sciences and that the objects they seek to study and manipulate have a special status. Indeed, exceptionalism continues to be ubiquitous in the debates surrounding these sciences. Specialists themselves are far from free of the tendency, and enthusiasts for their field are strikingly prone to it, but exceptionalist rhetoric is found in the work of their opponents as well; we have discussed several examples taken from the critical literature. In chapters 6 and 7, we looked at critical responses to the growth of the powers that have emerged from genomics and at arguments in favor of resisting or curtailing those powers. Many of these arguments, including both utilitarian criticisms citing risks and uncertainties and nonutilitarian criticisms citing threats to natural order and human dignity, could have been deployed against any number of targets. The new genetic/genomic technologies have been particularly vulnerable to specific exceptionalist versions of these arguments largely because of the exceptional status already associated with them.

We have already made clear our own reservations about critical arguments of this sort and rejected the suggestion that genomes and genomics are exceptional in any fundamental sense. Of course, in opposing exceptionalism of this kind we do not seek to criticize every casual reference to the "exceptional" features of genomes or genomics; just as with other objects or activities, it may sometimes make good pragmatic sense to point out exceptional aspects of them. And there is one entirely naturalistic sense in which genomics may very usefully be flagged up as exceptional. Arguably it is exceptional in its potency, both as science and as technology, and in the rate at which that potency is increasing.

Anxieties about the powers that genomics is threatening to unleash may inspire two kinds of criticism. First of all, because genomic powers almost inevitably fall into the possession of dominant institutions and organizations, they are an obvious target for anyone who regards those institutions and organizations as the enemy. The animus of the marginalized against the powerful is likely accordingly to inspire critical attacks upon genomics. Second, genomics and associated sciences may be attacked simply because, in themselves, they are power. There is a long tradition of criticism wherein the power of technology is seen as in itself a threat to what is good in the existing order of things, and a means for imposing ever-increasing control and creating a vast system of instrumentalities in which everyone is reduced to the status of means. Of course, all societies possess a technology and in practice it is only technological advance that is criticized in this way, and the latest and most potent technologies.

We shall be more sympathetic to the critics if we regard them not as opponents of technology as such, but as interrogators of power. How shall we live if and as the institutions and organizations that dominate our lives and the bureaucracies that are their means of doing so acquire more and more powers and capacities? And how shall we live with power, even if we ourselves count among the powerful or live in an equitable democracy, given that power is double-edged and that its constant proliferation can be a menace to those who wield it as well as those against whom it is exercised? These are large questions about the powers and institutions that are necessary for and/or compatible with a good life. Indeed, the second question, of whether and how we shall cope with our own ever-increasing powers, stands among the great unanswered questions of our times. It is entirely appropriate to dramatize this question by directing it at those powers currently being constituted and developed at the cutting edge of scientific research and technological innovation, of which the powers of genomics are currently preeminent examples. This is surely the main reason why genomics is now marked with an exceptional status and singled out for criticism. It is opposed as proliferating power, simply for being such power, and the arguments between critics and apologists serve to make us more and more aware of the awesome possibilities for good and ill that inhere in that power.

Genomics is widely perceived as being at the forefront of scientific and technological advance, and hence serves as a symbol for the ever-increasing potential, as well as the risk and threat, of new technology. In this regard it is given the role once assigned to nuclear physics. Nuclear technology was a technology of similarly daunting potency, also firmly located in the dominant institutions and organizations of modern societies, recognized both as power and as a symbol and manifestation of their power, and attacked with comparable intensity. And we can clearly see today how the question faced half a century ago of whether or not to continue to develop nuclear technologies was one with enduring implications for our lives. Unforeseen, or allegedly unforeseen, consequences of decisions taken then continue to unfold. Opportunities to use the technologies in new ways, including ways unimaginable fifty years ago, continue to present themselves. Pragmatic justifications for the rhetoric of nuclear exceptionalism so prominent at that time remain easy to find, and many people today ask whether there should not then have been an even more obsessive concern with the potential downside of that technology.

The nuclear science and technology of half a century ago is perhaps the only significant analogue of genomics as it stands today, but it is a very good one. Nuclear technology was singled out, as genomics now is, as conferring powers incomparably greater than any other emerging technology and engendering problems of a different order of magnitude. Here was a technology too awesome to use at all, in the hands of people all too likely to misuse it, or so millions of people believed. And increasing numbers are coming to the same view of genomic technologies. Even so, the analogy with nuclear power is less than perfect, and much can be learned from the differences it throws up as well as from the parallels it reveals. Even if the inspiration of the criticism was largely the same, nuclear technology was not criticized in quite the way that genomics is now criticized, and the exceptional status it was accorded was not rationalized in quite the same way. In particular, our responses to genomic powers are knowledge dependent to a degree that was never the case with nuclear technology. In part, this may be because genomics inspires many specific and detailed fears, whereas the power of nuclear technology inspired fear through its mere magnitude and its capacity to obliterate us. It may also be due in part to an increased interest in the details of technical issues, following the growth in their political salience that has been one of the major social changes of the last half century. In any event, whereas there was scarcely any connection between initial responses to nuclear technology and the contemporary knowledge of atoms and their nuclei, there is now an increasingly significant link between responses to genomics and knowledge of genes and genomes. And this exacerbates the need to encourage a properly naturalistic understanding of these things, and to combat those forms of exceptionalism that are inconsistent with it.

Accounting for exceptionalism

The genomic exceptionalism we are critical of here is not that which speaks of the exceptional magnitude of the powers of genomic technology or the exceptional risks and dangers of manipulations and interventions focused on genomes. Where "exceptional" refers only to degree or extent we would prefer not to speak of "exceptionalism" at all. What we identify as genomic exceptionalism is the assertion of a difference in kind between genomes and other components of organisms and an insistence that we treat them in a radically different way from other kinds of thing.

Genomic exceptionalism follows an earlier genetic exceptionalism and presently coexists with it. Both are associated with essentialist modes of thought. Just as our essential nature was previously reckoned to reside in our genes, so now there are references to our genome as our natural essence. And of course, while in principle it extends to all living things, it is *human* genes and genomes that are the predominant foci of this essentialist and hence exceptionalist discourse. In the previous chapter we criticized exceptionalist arguments directed against genomic manipulation by some philosophers, ethicists, and theologians and the misconceived genomic essentialism on which they appeared to rest. But the same criticisms should perhaps have been extended more widely; for the same essentialism appears increasingly to be permeating our entire culture. A very few human beings are genomic chimeras whose somatic cells derive from two different pairs of genomes and are of two different kinds—the presumption being that two fertilized eggs have fused together at an early stage of ontogeny.[1] The phenotypical consequences of chimerism vary from scarcely detectable to severe physiological problems, but none of them creates difficulties in establishing that a chimera is truly a single human being. Even so, this has not prevented the media from encouraging genomic essentialism by referring to "the twin within" such people, as if two persons are present in one body. The example could be used to illustrate what is wrong with genomic essentialism as it is currently encountered in the mass media, but we cite it here primarily as evidence of how widely genomic essentialism is diffusing. In this instance references to "the twin within" seem to have been made mainly in order to convey a simple understanding of what chimerism consists in. Recourse to genomic essentialism was made primarily in an effort to communicate with a general audience in a way that it would most readily

---

1 There are senses in which many human beings have a degree of genomic chimerism. Many people have undergone organ transplants or blood transfusions, and the variable suppression of the X chromosome in female bodies might be interpreted this way. Here we are talking of an extreme and rare variety of chimerism.

understand. But it also provides evidence of what those who are well placed to know believe a general audience would understand.[2]

Although genomic essentialism may be seen as a simple extension of the genetic essentialism already prominent in our cultural inheritance, this observation hardly suffices to explain either. We need to ask what underpins both these forms of essentialism. Why is our very nature as human beings widely regarded as residing in our genes, and now increasingly in our genomes, or as current essentialist discourse would have it, in "the human genome"? The only other material components of our bodies that we currently tend to regard in this way are our brains. While we view the excision, replacement, or modification of other parts of our bodies with relative equanimity, interference with brains or with genes/genomes may still elicit deep and widespread unease. Why is this?

The question of the nature of human nature that arises here is a famous one that has been formulated in many versions over the centuries, but these formulations did not initially presume a naturalistic understanding of human beings. The dominant problem in a long tradition of debate here was not primarily that of understanding the human body at all, but rather that of identifying the distinctive, morally vital features of human beings as participants in social life. And the solutions offered to it, solutions still recognized as salient today, were initially formulated long before genomes were known of or the role of the brain understood. What were these solutions? Needless to say, they varied; but the two putatively essential human characteristics that were identified again and again were reason and descent. If tradition is any guide, these are the elements, one a unique power of humans, the other a special set of relationships that links them to each other, that are essential in specifying what is human about human beings.[3]

---

2 Even *New Scientist*, in a useful review of chimerism of this sort (Ainsworth, 2003), nonetheless refers to "the stranger within" and describes a chimera as a "mixture of two individuals."

3 As well as reason and ancestry, currently reified, as we shall argue, as brain and genome, traditional accounts often cited free will as a feature of human nature; if the topic of interest here had been the ancient history of conceptions of what makes humans human then a discussion of free will would have been needed. Today, however, there is no material object that stands as a reification of free will. See Barnes (2000, 2002) and Dupré (2001, ch. 7) for accounts of free will that recognize that human beings are intrinsically social creatures, where Dupré treats it as a form of autonomy and Barnes identifies it as a specific kind of vulnerability.

The traditional approaches here have much to be said for them. Even when appraised from an anachronistically naturalistic standpoint, they have obvious merits. Human beings do stand out from other animal kinds by virtue of their possession of those various powers and competences we identify as "reason" or "rationality," which, crucially, include the power to rationalize actions and hence to establish mutual accountability and responsibility in relation to them. And many kinds of scientific evidence now link the existence of "reason" in humans to the functioning of something equally characteristic of them at the material level, their (relatively) large brains. Human beings can also be differentiated from other animal kinds by descent. They can be identified as a distinct interbreeding population, and their similarities can be rationalized by recourse to the traditional conception of "blood" as a material essence passed on down the generations. Scientific study now links the continuing existence of this interbreeding population to the recombination of human genomes that only get on well with genomes of their own kind, and it is possible now to regard genomes themselves as material analogues of "blood."

Traditional accounts of the uniquely valuable characteristics of human beings evolved over long periods in everyday contexts, not specialized settings, and everyday social interactions furnished most of the experience that engendered and gave credibility to their conceptions of human nature. Indeed, the "uniquely human" characteristics that feature so strongly in traditional accounts can be read precisely as specifications of the prerequisites necessary for social interaction. Successful social interaction requires that individuals possess appropriate powers and competences, and that they have some recognized position in a network through which other human beings may orient to them. The powers necessary for social interaction are very close to those traditionally associated with reason, and the social position to which others orient themselves was traditionally recognized as the status acquired by descent.[4] The associations proposed here may initially seem implausible in the context of our current strongly individualistic modes of thinking. We do not automatically connect reason with social accountability and responsibility, or even descent with social affiliation.[5] Yet it is not hard to reconstruct these connections and recognize their continuing significance. The performances that reveal a human to be possessed of

4 This recognition continues to this day in our use of surnames that denote relations of descent.

5 The derivation from the Latin, *filius*, is highly suggestive.

reason are still recognizable as those that establish her standing as a responsible, accountable social agent and parturition still establishes the key social relationships of kinship that define her initial place in a social network.

If reason and descent are intelligible as prerequisites for social interaction, then they are also the basis of the interactive, rather than purely instrumental, relations that exist between human beings. They permit humans to single each other out as potential foci of respect, and to treat each other as having dignity. And at this point the knot is tied with the conclusions of the discussion of human dignity in chapter 7. Those characteristics that in traditional accounts of "human nature" are "essentially human" turn out to be those that make humans worthy of respect. And if we have a preference for the latter formulation, it is easy nonetheless to forgive the essentialism of the traditional accounts; for it is hard to exaggerate the significance of the respect and dignity that human beings accord each other and the system of interaction in which it is accorded and that it underpins.

Many social scientists have observed how an orientation of respect and deference is a ubiquitous feature of social interactions and some have conjectured that the capacity to adopt this orientation is both a native endowment and a functional necessity in our lives as social creatures (Goffman, 1967).[6] In adopting it, members of collectives constitute in their interactions a system of social cooperation and control through which they are able to reduce conflict, initiate collective instrumental action, and create powers vastly greater than they could hope to engender as independent individuals (Barnes, 1988, 1995). As Tom Scheff (1988) has described it, collectives invariably constitute in their interactions a "universal bio-social system" of control he calls a "deference/emotion system." The basis of control lies in the need that all humans have for the respect conveyed to them in the implicit evaluations of others. They experience an intrinsically satisfying sense of pride when accorded it,

---

6 Goffman's findings also point to the importance of implicit nonverbal communication in communicating respect and acknowledging dignity in the course of interaction and thereby help to account for the curious fact that we know intuitively what dignity is and yet have great difficulty in putting it into words. While explicit verbal exchanges may carry more empirical information, we are inhibited from using them in everyday interaction to assign honor; implicit communication carries the most authentic signals of honor and respect. There are grounds for regarding this arrangement, in which face-to-face interaction carries two parallel streams of messages, as highly functional.

and a sense of shame when it is withdrawn. By virtue of this, all alike are vulnerable to those fellow members with whom they interact, whose evaluations imply respect and confirm their dignity; coordinated, collectively beneficial, and "moral" actions are encouraged in consequence. On this view of things, there is far more to dignity than either its traditional rendering as "elevation" or more recent individualistic accounts of it suggest. The conferral of dignity on members by members is actually the basis of all that collectives have managed to accomplish that independent individuals could not have, which is almost everything.

Let us return now to genome and brain. We have already noted how the latter can be seen as a material analogue of reason, and the former as a material analogue of ancestry (as previously was blood). And it is possible to go further still and claim that our brains are material repositories of reason, and our genomes material embodiments of our unique ancestry. Brain and genome can be regarded as materialist versions of reason and ancestry, through the study of which science may contribute to the ongoing project of understanding the nature of human nature. But while specialized scientific work might just possibly contribute to our self-understanding via this route, everyday understanding itself emphatically does not stand to benefit from transferring its interest in reason and ancestry to brain and genome. A historical concern with what is uniquely valuable in human beings, and on what basis they single each other out for special respect, will not be advanced by reflection confined to these material objects. If "genome" and "brain" are allowed to replace "descent" and "reason" in nonspecialized discourse, they will merely reify as objects what ought to be recognized as a relation and a power, and thereby open the way to misunderstandings. Indeed, they would be the reifications of reifications. The already less-than-satisfactory notions of "reason" and "descent" would be replaced with notions still more distant from, and still more liable to mask and obscure, the phenomena we would do better to address. And the misconceived view that the signs of our essential humanity are more readily discernible in molecular structures and neural networks than in what we do and how we act toward each other would be given unwarranted encouragement.

Almost a century ago now the sociologist Emile Durkheim (1915) sought to explain why certain peoples were given to totemic forms of worship, in which material objects were venerated and accorded a special respect. The ultimate object of veneration was actually society itself, Durkheim said. The existence and value of society was well understood intuitively, but it was not understood in the way that a material object may typically be understood, as something that can be seen and grasped.

So the totem stood in for society.[7] What was performatively constituted was replaced by something materially constituted. Perhaps there is an insight here into our continuing search for our material essence. Perhaps it is because what we understand as our essential nature is constituted performatively, in the context of social interaction, that some of us seek a more concrete grasp of it in the form of an object. Even if this speculation is set aside, it could still be that our inclination to identify the essence of ourselves with what is written in our genomes would greatly diminish if we chose to reflect more deeply on our social life, the interaction with others in which it consists, and the ways in which our dignity and standing are sustained therein by the respect we accord each other.

Life without essences: reduction as emancipation

In our efforts to make sense of the flux of experience, reification and essentialism may often be fruitful strategies.[8] But claims that our essential nature can be discovered in our genomes or our brains are instances where these strategies serve mostly to mislead. Whatever else, they are likely to serve us poorly in the context of everyday life. Given how references to "human nature" have functioned in that context, we would do better either to accept that there is no such thing and seek to make sense of the pattern of our lives in other ways, or, if we do wish to speak of such a thing, to identify human nature with our natural sociability and interactive capability.

Human beings are well described as naturally sociable and interactive. Even though they rarely describe each other explicitly in this way, they justify the description through their relations with each other. And the interactive capacities necessary for participation in social relations are repeatedly encountered in a long theoretical tradition as the distinctive characteristics that entitle humans to a special respect and dignity. Unfortunately, however, in the context of theory these capacities have tended to appear in reified forms, as objects or essences, and we ought to recognize the dangers to which the use of such props to the imagination gives rise and try to manage without them.

---

7 Durkheim himself has been criticized for reifying as "society" what the peoples he wrote of reified as a totem; in Durkheim "society" stands in for social interactions and social processes (Barnes, 1995).

8 Indeed, it has been argued that social interaction and the coordination of our behavior would not be possible without use of language involving reification (Thomason, 1982).

Sociability and social interaction are characteristic of developed human beings operating as competent social agents. They are not characteristics of genomes or zygotes or embryos or other supposed precursors of fully human entities, and are no more potentially present in these things than they are in the molecules or the gametes that are their precursors. It follows that the normal basis for the attribution of human dignity and respect is not present in genomes and analogous entities. And indeed our own view of the matter is that the dignity paradigmatically conferred upon humans competently functioning as social creatures should not be spread across all parts and components of the human life cycle. No conclusive case can be made for this view. Human dignity is something conferred, and if enough human beings confer it, then whatever it is conferred upon ipso facto possesses it. There is no formal argument that will expose the error in treating genomes, or zygotes, or animals, or body parts, or totem poles, or whatever else as possessed of human dignity. One can only point out how forced and tenuous is the analogy between these instances and the paradigm case, and how it is impossible in practice to treat them as the paradigm case is treated.

On the matter of human dignity, then, our view is close to that of Jürgen Habermas that we discussed, perhaps a little too critically, in the previous chapter. Human dignity should not be attributed to human genomes, and we should reject criticisms of genomics and genomic manipulation that specifically depend on such an attribution.[9] Our concern, however, has not been merely to rebut criticisms of this sort, but rather to call into question essentialist and exceptionalist accounts of genomes wherever they appear, whether in critical attacks on genomics or in its hagiography. It is worth recalling here the way that advances in the new molecular genetics and eventually in genomics were initially presented to wider audiences. The structure of DNA may or may not have been proclaimed the secret of life in 1953, but it was again and again in the 1970s and '80s. Chapter 2 briefly alluded to how the DNA molecule was characterized as exceptional: it was the template for a complete human being; a set of instructions; a code; a program; the text of the book of life; the master molecule. It is scarcely surprising in the light of this that commentators even now continue to describe "the human genome" as our natural essence, but, again as we described in chapter 2, this DNA fetishism also elicited a critical response.

9 Recall that it is not human dignity but only "the dignity of human life," a different and a lesser dignity, that Habermas is willing to see accorded to genomes.

DNA was said to be just one ingredient in cell chemistry, no more significant causally than salt or water, essential only in the way that they too are essential. The standing of DNA as the essential uncaused cause that set in motion a great chain of effects, its position at the pinnacle of a hierarchy of control, was a mythological reconstruction of a complex interactive biological system better regarded as analogous to a democratic polity. In general, DNA was no big deal. We have already discussed the limitations of this kind of claim. It was rather like claiming that the plutonium in the bomb was no big deal, being no more important causally than the detonator and the rest of the bomb's components. Like the plutonium in the bomb, the DNA in the cell is a big deal, although it can be curiously difficult to say just why, and the metaphors deployed by enthusiasts to try to convey why have been easy meat for critical philosophers seeking to deflate and undermine what they have seen as pretentious rhetoric.

In the to and fro of debate, the use of an essentialist and exceptionalist rhetoric is likely to prove almost irresistibly tempting as the need arises to highlight the importance of something, whether in order to celebrate its value or to expose the threat it represents. By pricking the bubble of essentialist and exceptionalist rhetoric surrounding DNA, critics of genomics sought to deflate the overblown pretensions of the associated science. If DNA is exceptional, then so is the science that studies it; if DNA is no big deal, then neither is the science that studies it. But now that the science has become a big deal notwithstanding, and its power and perceived importance have grown, deflationary rhetoric may be counterproductive. Now the critic may wish to expose and even exaggerate the threat that the power of genomics represents, and be drawn back to the rhetoric of genomic essentialism and exceptionalism both to point up the unique character of the threat and the way to set bounds upon it. And indeed there does seem to have been a significant shift in the use of essentialist rhetoric over time from enthusiasts to critics of genomics, which may tell us something of the power that genomics is continuing to accumulate.

The politics of science, both its internal bureaucratic politics and the politics of its reception and assimilation, are always fought out, in part, as a competition between different descriptions of scientific objects and different assessments of their ontological status. In the specific context of genomics, both exceptionalist and naturalistic accounts of genomes currently feature among these descriptions. And argument rages about whether genomes are carriers of the natural essences of organisms, or variable bits of DNA carried by organisms, or theoretical constructions used to understand organisms, or repositories of information about organisms stored in computer memories, or mere figments of the human imagination. Indeed, it is not possible to find an existing way of describing what genomes are that is not associated with one side or other in some such debate, which raises the question of how "genomes themselves" are best described in a book such as this one. There is no fully satisfactory answer to this question. We do not claim that the descriptive framework we have actually employed is the correct one or even the best one. It does, however, allow us to give clear expression to the naturalism we both espouse and regard as characteristic of the sciences. And it is a choice strongly influenced by recent developments in the science itself.

The decline in the standing of the gene concept, first discussed in chapter 2, together with the shift of ontological authority to concepts drawn from chemistry, has been at once a scientific necessity and a stroke of good fortune for us. The gene concept has inspired many remarkable scientific achievements over the years and is sure to continue to do so. But as it is moved to the sidelines ontologically, all kinds of unwanted associations can be sidelined with it, and in particular those most harmful in the context of everyday understanding, in which the discourse of genetic essentialism has long been entangled with every sort of nastiness. Instead, it is now possible to turn to the concepts of molecular stereochemistry that are currently authoritative ontologically and crucial theoretically in genomics but still comparatively little stained by the crudest forms of political and ideological misuse. It is these concepts that we have made the basis of our own account. Genomes are really molecules, we say, and like all molecules they are material objects. Of course, this way of solving our problem is open to challenge. An obvious objection is that in its acceptance of current ontological orthodoxy our account abandons a critical perspective on genomics and connives in its project of reducing biology to physics and chemistry. But against this we would simply say that if endorsement of the move away from gene realism and genetic essentialism is reductionism, roll on reductionism. In fact, of course, as we have noted in earlier chapters, we do not regard the move from gene realism to the structural/holistic schema of molecular stereochemistry as reductionist at all.

In brief, we take genomes to be material objects. This implies nothing, of course, about what "the" human genome might be, or indeed what "the" genome of any species might be. Our assumption concerns genomes as particulars and merely mirrors what specialist researchers themselves typically assume in the course of their practice. For specialists, the idea that genomes are really complex, spatially extended chemical molecules is now a fertile source of conjectures, and an inspiration for research in which they can have confidence. But in the context of everyday understanding, the same idea, which effectively characterizes genomes as mundane, composite material objects, is also valuable as a means of emancipation from the grip of misleading notions and unwanted associations. Crucially, in being clear that genomes are in themselves no more than bits of material, we are less likely to mistake any additional status or standing we impute to genomes for an intrinsic characteristic of the genomes themselves. The status becomes easier to recognize as an expression of our decision to treat the object in a given way.

The distinction between the nature of objects and the status we assign to them, between how they are and how we treat them, tends to be harder to make in proportion to the importance of making it. It is especially hard to make when we are not able to agree on how to treat the objects in question. Then the temptation is constantly to refer back to the objects, whether in search of further knowledge or of excuses for agreements that have simply got to be made come what may. Zygotic genomes are objects of this sort. Currently, they lie close to a key cultural boundary, with gametes, to which we are generally indifferent, on their left, embryos, which increasingly concern us morally as they develop, on their right, and a major shift of orientation having to be made and justified at some point between the two. The precise point at which this shift should be made is evidently something on which we cannot agree. But it is hard to see how further knowledge of genomes as physical objects could be of assistance here, and hence how genomics could be relevant when reappraising and perhaps deciding between the different statuses currently assigned to these problematic entities. We need to look to ourselves here, not to our genomes.

A critic might be tempted to describe this book as an attempt to replace old outmoded myths with a new materialist mythology of genomics. Certainly our account is resolutely materialistic and may be said to presuppose a materialist ontology. As we have already noted, one significant virtue of materialism here is that it highlights the distinction between claims about the status of genomes and claims about their nature and makes us more critically aware of the innumerable instances where the two things are confused or the one is disguised as the other. Provided only that they are not so disguised, however, accounts of the status merited by genomes, or even of the exceptional status that they ought to be accorded, need not clash with our own account. It hardly needs saying, for example, that there are good reasons for treating some genomes with exceptional caution and restraint, material objects though they are. They are reasons that reflect our awareness of the dangers that attend the use of power and our consequent ambivalence in the face of power. The problem is that even here we remain liable to confuse the nature of things with how we treat them. We continue to speak of the possibilities and the dangers of altering the master molecule that controls so much of our development and affects its outcome so profoundly. But our hopes and fears are actually focused on ourselves and elicited by our own new powers to act, whether for good or ill. No doubt we are well aware of this at one level, but our discourse and its favored terminology have turned things upside down. DNA gives cause for concern as a mastered molecule, not as a master molecule.

Bibliography

Aderem, A. 2005. Systems biology: its practice and challenges. *Cell* 121:511–513.

Ainsworth, C. 2003. The stranger within. *New Scientist* 2421:34–37.

Altamura, S., P. Cammarano, and P. Londei. 1986. Archaeobacterial and eukaryotic ribosomal subunits can form active hybrid ribosomes. *FEBS Lett.* 204:129–133.

Ariew, A. 1996. Innateness and canalization. *Philosophy of Science* 63:S19–S27.

Arthur, W. B. 1984. Competing technologies and economic prediction. *Options* 2:10–13.

Avery, O. T., C. M. MacLeod, and M. McCarty. 1944. Studies on the chemical nature of the substance inducing transformation of pneumococcal types. *Journal of Experimental Medicine* 79:137–158.

Banton, M. 1983. *Racial and Ethnic Competition*. Cambridge: Cambridge University Press.

Barnes, B. 1988. *The Nature of Power*. Cambridge: Polity Press.

Barnes, B. 1995. *The Elements of Social Theory*. London: UCL Press.

Barnes, B. 2000. *Understanding Agency*. London: Sage.

Barnes, B. 2002. The public evaluation of science and technology. In J. A. Bryant et al. (eds.), *Bioethics for Scientists*. Chichester: John Wiley.

Barnes, B., et al. 1996. *Scientific Knowledge: A Sociological Analysis*. Chicago: University of Chicago Press.

Bassler, B. L. 2002. Small talk: cell-to-cell communication in bacteria. *Cell* 109:421–424.

Beck, U. 1988. *Risk Society: Towards a New Modernity*. London: Sage.

Beck, U. 1992. From industrial society to the risk society: questions for survival, social structure, and ecological enlightenment. *Theory, Culture, and Society* 9:97–123.

Bedau, M. A. 2003. Artificial life: organization, adaptation, and complexity from the bottom up. *Trends in Cognitive Science* 7:505–512.

Bejerano, G., et al. 2004. Ultraconserved elements in the human genome. *Science* 304:1321–1325.

Bendich, A. J., and K. Drlica. 2000. Prokaryotic and eukaryotic chromosomes: what's the difference? *BioEssays* 22:481–486.

Beurton, P., R. Falk, and H.-J. Rheinberger (eds.). 2000. *The Concept of the Gene in Development and Evolution.* Cambridge: Cambridge University Press.

Bostanci, A. 2008. *The Human Genome: Something We All Share.* Ph.D. thesis, University of Exeter.

Boyle, E. 2004. After the Millennium. *Nature Genetics,* vol. 36 supplement.

Brown, T. A. 2002. *Genomes.* Oxford: BIOS.

Cherkas, L. F., E. C. Oelsner, Y. T. Mak, A. Valdes, and T. D. Spector. 2004. Genetic influences on female infidelity and number of sexual partners in humans. *Twin Research* 7:649–658.

Cooke, B., and U. Kotheri (eds.). 2001. *Participation: The New Tyranny.* London: Zed Books.

Cooper, S. J., G. A. Leonard, S. M. McSweeney, A. W. Thompson, J. H. Naismith, S. Qamar, A. Plater, A. Berry, and W. N. Hunter. 1996. The crystal structure of a class II fructose-1,6-bisphosphate aldolase shows a novel binuclear metal-binding active site embedded in a familiar fold. *Structure* 4:1303–1315.

Crick, F. H. C. 1958. On protein synthesis. *Symposia of the Society for Experimental Biology* 12:139–163.

Darnell, J., H. Lodish, and D. Baltimore. 1990. *Molecular Cell Biology,* 2nd ed. New York: Freeman.

Daston, L., ed. 2000. *Biographies of Scientific Objects.* Chicago: University of Chicago Press.

Dawkins, R. 1976. *The Selfish Gene.* Oxford: Oxford University Press.

Dobzhansky, T. 1973. Nothing in biology makes sense except in the light of evolution. *American Biology Teacher* 35:125–129.

Doolittle, W. F. 2005. If the tree of life fell, would we recognize the sound? In J. Sapp (ed.), *Microbial Phylogeny and Evolution: Concepts and Controversies.* Oxford: Oxford University Press.

Douglas, M. 1973. *Natural Symbols: Explorations in Cosmology.* London: Routledge.

Douglas, M. 1975. *Implicit Meanings: Essays in Anthropology.* London: Routledge.

Douglas, M. 1966. *Purity and Danger.* London: Routledge.

Douglas, M. 1986. *Risk Acceptability According to the Social Sciences.* New York: Russell Sage Foundation.

Douglas, M. 1992. *Risk and Blame.* London: Routledge.

Douglas, M., and D. Hull (eds.) 1992. *How Classification Works: Nelson Goodman among the Social Sciences.* Edinburgh: Edinburgh University Press.

Douglas, M., and Wildavsky, A. 1982. *Risk and Culture.* Berkeley: University of California Press.

Dupré, J. 1993. *The Disorder of Things: Metaphysical Foundations of the Disunity of Science*. Cambridge, Mass.: Harvard University Press.

Dupré, J. 2001. *Human Nature and the Limits of Science*. Oxford: Oxford University Press.

Dupré, J. 2002. *Humans and Other Animals*. Oxford: Oxford University Press.

Dupré, J. 2003. *Darwin's Legacy: What Evolution Means Today*. Oxford: Oxford University Press.

Dupré, J. 2004. Understanding contemporary genomics. *Perspectives on Science* 12:320–338.

Dupré, J. 2005. Are there genes? In A. O'Hear (ed.), *Philosophy, Biology and Life*. Cambridge: Cambridge University Press.

Dupré, J. 2007. The inseparability of science and values. In H. Kincaid, J. Dupré, and A. Wylie (eds.), *Value-Free Science: Ideal or Illusion*. New York: Oxford University Press.

Dupré, J. 2008. What genes are, and why there are no "genes for race." In Barbara A. Koenig, Sandra Soo-Jin Lee, and Sarah Richardson (eds.), *Revisiting Race in a Genomic Age*. New Brunswick: Rutgers University Press.

Dupré, J., and M. O'Malley. 2007. Metagenomics and biological ontology. *Studies in the History and Philosophy of the Biological and Biomedical Sciences* 38: 834–846.

Durkheim, E. [1915] 1976. *The Elementary Forms of Religious Life*, 2nd ed. London: Unwin.

Ellis, B. 2001. *Scientific Essentialism*. Cambridge: Cambridge University Press

Epstein, S. 1995. The construction of lay expertise: AIDS activism and the forging of credibility in the reform of clinical trials, *Science, Technology, & Human Values* 20:408–437.

Epstein, S. 1996. *Impure Science: AIDS, Activism, and the Politics of Knowledge*. Berkeley: University of California Press.

Fogle, T. 2000. The dissolution of protein coding genes in molecular biology. In Beurton, Falk, and Rheinberger 2000.

Forterre, P. 2006. Three RNA cells for ribosomal lineages and three DNA viruses to replicate their genomes: a hypothesis for the origin of cellular domains. *PNAS* 103:3669–3674

Foucault, M. 1977. *Discipline and Punish*. New York: Vintage.

Foucault, M. 1980. *Power/Knowledge*. New York: Pantheon.

Fox Keller, E. 1994. Master molecules. In C. F. Cranor (ed.), *Are Genes Us?* New Brunswick: Rutgers University Press.

Fox Keller, E. 2000. *The Century of the Gene*. Cambridge, Mass.: Harvard University press.

Galperin, M. Y. 2004. Metagenomics: from acid mine to shining sea. *Environmental Microbiology* 6:543–545.

Galperin, M. Y., D. R. Walker, and E. V. Koonin. 1998. Analogous enzymes: independent inventions in enzyme evolution. *Genome Research* 8:779–790.

Goffman, E. 1967. *Interaction Ritual: Essays on Face-to-Face Behavior*. New York: Doubleday.

Gogarten, J. P., and J. P. Townsend. 2005. Horizontal gene transfer, genome innovation and evolution. *Nature Reviews Microbiology* 3:679–687.

Goodman, N. 1954. *Fact, Fiction, and Forecast*. London: University of London Press.

Gregory, T. R. (ed.). 2004. *The Evolution of the Genome*. Oxford: Academic.

Griffiths, P., and R. Gray. 1994. Developmental systems and evolutionary explanation. *Journal of Philosophy* 91:277–304.

Griffiths, P., and K. Stotz. 2006. Genes in the post-genomic era. *Theoretical Medicine and Bioethics* 27:499–521.

Habermas, J. 2003. *The Future of Human Nature*. Cambridge: Polity.

Hauskeller, C. 2004. Genes, genomes, and identity: projections on matter. *New Genetics and Society* 23:285–299.

Hauskeller, C. 2005. Science in touch: functions of biomedical terminology. *Biology and Philosophy* 20:815–835.

Heidegger, M. 1977. *The Question Concerning Technology and Other Essays*. Trans. W. Lovett. New York: Harper and Row.

Haig, D. 2002. *Genomic Imprinting and Kinship*. Rutgers University Press: New Brunswick.

Hershey, A. D., and M. Chase. 1952. Independent functions of viral protein and nucleic acid in growth of bacteriophage. *Journal of General Physiology* 36:39–56.

Holme, I. 2007. *Genetic Sex: A Symbolic Struggle Against Reality*. Ph.D. thesis, Exeter University.

Hughes, S. 2005. Navigating genomes: the space in which genes happen. *Tailoring Biotechnologies* 1:35–46.

Hull, D. 1992. Biological species. In Douglas and Hull 1992.

Ideker, T., V. Thorsson, J. A. Ranish, R. Christmas, J. Buhler, J. K. Eng, R. Bumgarner, D. R. Goodlett, R. Aebersold, and L. Hood. 2001. Integrated genomic and proteomic analyses of a systematically perturbed metabolic network. *Science* 292: 929–934.

Inglehart, R. 1997. *The Silent Revolution*. Princeton: Princeton University Press.

International Theological Commission. 2004. *Communion and Stewardship: Human Persons Conceived in the Image of God* (Vatican Statement on Creation and Evolution). Vatican City: International Theological Commission.

Jablonka, E., and M. Lamb. 2005. *Evolution in Four Dimensions: Genetic, Epigenetic, Behavioral, and Symbolic Variation in the History of Life*. Cambridge, Mass.: MIT Press/Bradford Books.

Jacob, F., and J. Monod. 1961. Genetic regulatory mechanisms in the synthesis of proteins. *Journal of Molecular Biology* 3:318–356.

Jasanoff, S. 2003. Accounting for expertise. *Science and Public Policy*. 30: 157–162.

Jasanoff, S. 2004. *States of Knowledge: The Co-Production of Science and Social Order*. London: Routledge.

Jasanoff, S. 2005. *Designs on Nature: Science and Democracy in Europe and the United States*. Oxford: Princeton University Press.

Jones, S. 1994. *The Language of the Genes*. London: Flamingo Press.

Jones, S. 1996. *In the Blood: God, Genes, and Destiny*. London: Harper Collins.

Keasling, J. 2007. Video Lecture on Synthetic Biology. http://media.coe.berkeley.edu/ BIOSECURITY/03082007/BioLC8.asx

Kevles, D. 1985. *In the Name of Eugenics*. London: Penguin.

Kimura, M. 1983. *The Neutral Theory of Molecular Evolution*. Cambridge: Cambridge University Press.

Kincaid, H., J. Dupré, and A. Wylie (eds.). 2007. *Value-Free Science: Ideal or Illusion?* New York: Oxford University Press.

Kitcher, P. 1996. *The Lives to Come*. New York: Simon and Schuster.

Kohler, R. 1994. *Lords of the Fly: Drosophila Genetics and the Experimental Life*. Chicago: University of Chicago Press.

Kuhn, T. S. 1970. *The Structure of Scientific Revolutions*, 2nd ed. Chicago: University of Chicago Press.

Kuhn, T. S. 1977. Second thoughts on paradigms. In *The Essential Tension: Studies in Scientific Tradition and Change*. Chicago: University of Chicago Press.

Lewontin, R. C. 1991. *Biology as Ideology*. Concord, Ontario: Anansi.

Mackie, J. L. 1974. *The Cement of the Universe*. Oxford: Clarendon Press.

Marks, J. 2002. *What It Means to Be 98% Chimpanzee: Apes, People, and Their Genes*. Berkeley: University of California Press.

Marshall, E. 2005. Will DNA bar codes breathe life into classification? *Science* 307: 1037.

McCarroll, S. A., et al. 2006. Common deletion polymorphisms in the human genome. *Nature Genetics* 38:86–92.

M'charek, A. 2005. *The Human Genome Diversity Project*. Cambridge: Cambridge University Press.

Migeon, B. R. 2007. *Females Are Mosaics*. Oxford: Oxford University Press.

Minkel, J. R. 2006. Human-chimp gene gap widens from tally of duplicate genes. *Scientific American*. December, 19.

Moss, L. 2003. *What Genes Can't Do*. Cambridge, Mass.: MIT Press/Bradford Books.

Müller-Wille, S. 2005. Early Mendelism and the subversion of taxonomy: epistemological obstacles and institutions. *Studies in the History and Philosophy of Biological and Biomedical Sciences* 36:465–487.

Müller-Wille, S., and H.-J. Rheinberger (ed.), 2006. *Heredity Produced: At the Crossroads of Biology, Politics and Culture, 1500–1870*. Cambridge: MIT Press.

Mutz, D. C. 2006. *Hearing the Other Side*. Cambridge: Cambridge University Press.

Nature Genetics. 2004. Genetics for the Human Race. Vol. 36, supp.

Nelkin, D. (ed.). 1979. *Controversy: Politics of Technical Decisions*. London: Sage.

Nelkin, D. 1975. The political impact of technical expertise. *Social Studies of Science* 5:35–54.

Nelkin, D., and L. Tancredi. 1989. *Dangerous Diagnostics: The Social Power of Biological Information*. Chicago: University of Chicago Press.

Nickles, T. 2003. *Thomas Kuhn*. Cambridge: Cambridge University Press.

Noble, D. 2006. *The Music of Life*. Oxford: Oxford University Press.

Nowotney, H. 2003. Democratising expertise and socially robust knowledge. *Science and Public Policy* 30 (June): 151–156.

O'Malley, M., and J. Dupré. 2005. Fundamental issues in systems biology. *BioEssays* 27:1270–1276.

O'Malley, M., and J. Dupré. 2007. Size doesn't matter: towards a more inclusive philosophy of biology. *Biology and Philosophy* 22:155–191.

Odling-Smee, F. J., K. Laland, and M. W. Feldman. 2003. *Niche Construction: The Neglected Process in Evolution.* Princeton: Princeton University Press.

Oppenheimer, J. R. 1958. The tree of knowledge. *Harper's* October, 55–60.

Oyama, S. 2000. *The Ontogeny of Information: Developmental Systems and Evolution,* 2nd ed. Durham, NC: Duke University Press.

Paley, W. 1802. *Natural Theology: or, Evidences of the Existence and Attributes of the Deity, Collected from the Appearances of Nature.*

Pearson, H. 2006. Genetics: what is a gene? *Nature* 441:398–401.

Pellizzoni, L. 2003. Uncertainty and participatory democracy. *Environmental Values* 12:195–224.

Pembrey, M., L. O. Bygren, G. P. Kaati, et al. 2005. Sex-specific, sperm-mediated transgenerational responses in humans. *European Journal of Human Genetics* 14:159–166.

Pennisi, E. 2007. Jumping genes hop into the evolutionary limelight. *Science* 317:894–895.

Pinker, S. 2002. *The Blank Slate: The Modern Denial of Human Nature.* New York: Penguin.

Quine, W. V. O. 1977. Natural kinds. In *Ontological Relativity and Other Essays.* New York: Columbia University Press.

Rabinow, P. 1996. *Making PCR: A Story of Biotechnology.* Chicago: University of Chicago Press.

Rabinow, P., and N. Rose. 2003. *Biopower Today.* London: London School of Economics.

Radman M., I. Matic, and F. Taddei. 1999. The evolution of evolvability. *Annals of the New York Academy of Science* 870:146–155.

Rawls, J. F., B. S. Samuel, and J. I. Gordon. 2004. Gnotobiotic zebrafish reveal evolutionarily conserved responses to the gut microbiota. *PNAS* 101:4596–4601.

Rayner, S. 2003. Democracy in the age of assessment: reflections on the roles of expertise and democracy in public-sector decision making. *Science and Public Policy* 30:163–170.

Rayner, S., and E. Malone 1998. *Human Choice and Climate Change.* Washington: Batelle Institute Press.

Reardon, J. 2005. *Race to the Finish.* Princeton: Princeton University Press.

Redon, R., et al. 2006. Global variation in copy number in the human genome. *Nature* 444:444–454.

Renault, N., et al. 2007. Heritable skewed X-chromosome inactivation leads to haemophilia A expression in heterozygous females. *European Journal of Human Genetics* 15:628–637.

Rheinberger, H.-J. 1997. *Toward a History of Epistemic Things.* Stanford: Stanford University Press.

Ridley, M. 1999. *Genome.* London: Harper and Row.

Robert, J. S. 2004. *Embryology, Epigenesis and Evolution: Taking Development Seriously.* Cambridge: Cambridge University Press.

Rose, N. 2007. *The Politics of Life Itself.* New Jersey: Princeton University Press.

Roselló-Mora, R. and R. Amann. 2001. The species concept for prokaryotes. *FEMS Microbiology Reviews* 25: 36–67.

Rowe, G., and L. J. Frewer. 2004 Evaluating public-participation exercises: a research agenda. *Science, Technology, & Human Values* 29:512–556.

Saunders, N. J., P. Boonmee, J. F. Peden, and S. A. Jarvis. 2005. Inter-species horizontal transfer resulting in core-genome and niche-adaptive variation within *Helicobacter pylori*. *BMC Genomics* 6:9. http://www.pubmedcentral.nih.gov/articlerender.fcgi?artid=549213.

Savage, D. C. 1977. Microbial ecology of the gastrointestinal tract. *Annual Review of Microbiology* 31:107–133.

Scheff, T. J. 1988. Shame and conformity: the deference-emotion system. *American Sociological Review* 53:395–406.

Schwartz, J. 2005. *The Red Ape: Orangutans and Human Origins*. Rev. and updated ed. Boulder: Westview Press.

Shannon, C., and W. Weaver. 1948. *A Mathematical Theory of Communication*. Urbana: University of Illinois Press.

Shapiro, J. A. 2005. A 21st century view of evolution: genome system architecture, repetitive DNA, and natural genetic engineering. *Gene* 345:91–100.

Shapin, S., and S. Schaffer. 1985. *Leviathan and the Air Pump*. Princeton: Princeton University Press.

Sherrington, C. 1940. *Man on His Nature*. Cambridge: Cambridge University Press.

Silver, L. M. 1998. *Remaking Eden*. New York: Harper.

Singh, D. 1993. Adaptive significance of waist-to-hip ratio and female physical attractiveness. *Journal of Personality and Social Psychology* 65:293–307.

Sober, E. 1988. *Reconstructing the Past: Parsimony, Evolution, and Inference*. Cambridge, Mass.: MIT Press/Bradford Books.

Sokal, R. R., and P. H. A. Sneath. 1963. *Principles of Numerical Taxonomy*. San Francisco: W. H. Freeman.

Stotz, K., and P. E. Griffiths. 2004. Genes: philosophical analyses put to the test. *History and Philosophy of the Life Sciences* 26:5–28.

Stotz, K., P. E. Griffiths, and R. Knight. 2004. How scientists conceptualise genes: an empirical study. *Studies in History & Philosophy of Biological and Biomedical Sciences* 35:647–673.

Sudbery, P. 2002. *Human Molecular Genetics*. Englewood Cliffs, N.J.: Prentice Hall.

Sulston, J., and G. Ferry. 2002. *The Common Thread*. London: Bantam Press.

Suttle, C. A. 2005. Viruses in the sea. *Nature* 437:356–361.

Suttle, C. A. 1994. The significance of viruses to mortality in aquatic microbial communities. *Microbial Ecology* 28:237–243

Tauber, A. I., and Sarkar, S. 1992. The human genome project: has blind reductionism gone too far? *Perspectives on Biology and Medicine* 35:220–235.

Thomason, B.C. 1982. *Making Sense of Reification*. Atlantic Highlands, N.J.: Humanities Press.

Venter, J. C. 2007. *A Life Decoded: My Genome, My Life*. Allen Lane: London.

Venter, J. C., K. Remington, J. F. Heidelberg, et al. 2004. Environmental genome shotgun sequencing of the Sargasso Sea. *Science* 304:66–74.

Villarreal, L. P. 2004. Can viruses make us human?. *Proceedings of the American Philosophical Society* 148:296–323.

Vines, G. 1998. Hidden inheritance. *New Scientist* 2162:27–30.

Warburg, O. and W. Christian. 1943. Isolierung und Kristallization des Garungsferments Zymohexase. *Biochem. Z.* 314:149–176.

Waters, C. K. 2004. What was classical genetics? *Studies in the History and Philosophy of Science* 35:783–809.

Welch, E. M., et al. 2007. PTC124 targets genetic disorders caused by nonsense mutations. *Nature* 447:87–91.

West-Eberhard, M. J. 2003. *Developmental Plasticity and Evolution*. New York: Oxford University Press.

Wildman, D. E., et al. 2003. Implications of natural selection in shaping 99.4% non-synonymous DNA identity between humans and chimpanzees. *PNAS* 100:7181–7188.

Wilsdon, J., and R. Willis. 2004. *See-Through Science: Why Public Engagement Needs to Move Upstream*. London: Demos.

Wilson, E. O. 1975. *Sociobiology*. Cambridge, Mass.: Harvard University Press.

Woese, C. R. 1987. Bacterial evolution. *Microbiological Review* 51: 221–271.

Woese, C. R., and G. G. Fox. 1977. Phylogenetic structure of the prokaryotic domain: the primary kingdoms. *PNAS* 11:5088–5090.

Woodward, J. 2003. *Making Things Happen: A Theory of Causal Explanation*. New York: Oxford University Press.

Wynne, B. 1989. Sheepfarming after Chernobyl: a case study in communicating scientific information. *Environment* 31: 10–15, 33–39.

Yu, D. W., and G. H. Shepard, Jr. 1998. Is beauty in the eye of the beholder? *Nature* 396:321–322.

Zald, M., and J. McCarthy. 1987. *Social Movements in an Organisational Society*. New Brunswick, N.J.: Transaction Books.

Zuckerkandl, E., and L. Pauling. 1965. Molecules as documents of evolutionary history. *Journal of Theoretical Biology* 8:357–366.

Index

DATE DUE

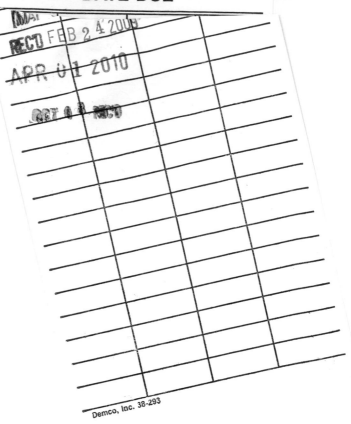

Demco, Inc. 38-293